OXFORD STATISTICAL SCIENCE SERIES

SERIES EDITORS

A. C. ATKINSON   J. B. COPAS
D. A. PIERCE   M. J. SCHERVISH
D. M. TITTERINGTON

OXFORD STATISTICAL SCIENCE SERIES

Practical Methods for Reliability Data Analysis

J. I. ANSELL

University of Edinburgh

and

M. J. PHILLIPS

University of Leicester

CLARENDON PRESS · OXFORD

1994

Oxford University Press, Walton Street, Oxford OX2 6DP

Oxford New York

Athens Auckland Bangkok Bombay
Calcutta Cape Town Dar es Salaam Delhi
Florence Hong Kong Istanbul Karachi
Kuala Lumpur Madras Madrid Melbourne
Mexico City Nairobi Paris Singapore
Taipei Tokyo Toronto

and associated companies in
Berlin Ibadan

Oxford is a trade mark of Oxford University Press

Published in the United States by
Oxford University Press Inc., New York

A catalogue record for this book is available from the British Library

Library of Congress Cataloging in Publication Data available

ISBN 0 19 853664 X

Typeset by the authors
Printed in Great Britain on acid-free paper by
Bookcraft (Bath) Ltd, Midsomer Norton, Avon

To Catherine
and
Carol, Emma, and Rachel

Preface

The study of reliability engineering is concerned with the improvement, analysis, assessment, and prediction of the performance of equipment and systems. The aim of a reliability study is to gain the best performance within the resources available. Achieving a high reliability in the performance of equipment is of utmost importance. For safety systems the ever increasing demands from regulatory bodies force engineers to establish higher levels of performance. In commercial contexts producers have to satisfy customers' requirements for equipment performance in order to retain market competitiveness.

Part of the process of achieving these goals of high reliability is the analysis of reliability data. Many reliability analyses require the use of statistical techniques to aid in the analysis, assessment, and prediction of the performance of equipment and systems. It is these problems which this book addresses. This book tackles the analysis of reliability data from a practical standpoint, giving a critical appreciation of the statistical methods and models used. The aim is to provide practitioners with guidance on the use of the statistical techniques. An important feature of the book is the setting of the context and tackling issues which reliability engineers actually face. It is important to appreciate where the data arise from and how they have been collected. There is a need to comprehend the background and nature of the data before they are analysed. Statistical theory has been kept to a minimum with the inclusion of only the material we believe is necessary.

We also hope the book will provide statisticians with an insight into the practical aspects of reliability engineering and some of the issues reliability engineers face. Often when reliability methods are taught to statisticians the prime interest has been on fitting lifetime distributions rather than on tackling the specific queries of reliability engineers. In most cases the reliability engineer wishes to assess whether changes have made an improvement or to assess whether a target reliability has been achieved by a system. Throughout the text we illustrate the methodologies by using examples obtained from our own consultancy experience or from the reliability literature.

In Chapter 1 we address the nature of reliability data, giving a range of examples including three case studies which are considered in detail in Chapter 10. We also discuss the sources of data such as historic databases and the construction of databases for reliability studies. The final part of the chapter provides the necessary probability background for the rest of the text. Chapter 2 considers lifetime distributions, both parametric and non-parametric, together with the issues of censoring and truncation, which are of utmost importance in reliability studies. Details of estimation procedures are given and inferential methods discussed. As data collection has improved, information is often available on other measurements besides times of failure or lifetimes. The use of this information, usually referred to as the covariates, is treated in Chapter 3, in which a series of models are explored.

Often in reliability studies an attempt is made to predict the performance of large and complex systems which are composed of many components. In Chapter 4 we consider the basic ideas for analysing the reliability of systems constructed from a number of components (or subsystems), and introduce the concept of the structure function and its representation. Also in this chapter approximations are considered as well as the specific problems of assessing and optimizing the performance of systems, when the systems may be subject to two kinds of failure. The latter part of the chapter is devoted to obtaining a representation of the system. Obviously many systems are composed of components which on failure can be repaired. In Chapters 5 and 6 we discuss such systems and the analyses associated with them. In Chapter 5 the stochastic models for repairable systems are examined and in Chapter 6 we explore the analysis of data arising from such systems.

Given that a major aspect of a reliability engineer's role is concerned with the improvement of equipment and systems, Chapter 7 critically reviews reliability growth models. As systems have become larger and more complex, dependency has been seen to play a role in the reliability of systems. In Chapter 8 we discuss the various models from engineering and statistics which are used for analysing dependency.

With the emphasis on practical methods for data analysis in this book, we felt it necessary to explore in Chapter 9 a series of techniques which would aid reliability engineers in initially understanding the nature of their data. In Chapter 10 we explore three case studies illustrating the types of problems and analyses faced by reliability engineers. We conclude with an Appendix which includes examples of S-PLUS and SAS programs to produce residual plots.

We are grateful to Elsevier Science Limited, Pergamon Imprint, Oxford, England, for permission to reprint Table 1.7 from *Reliability Engineering and Safety System*, **34**, 7–21, Newton, D.A., Some pitfalls in reliability data analysis, (1991), Table 3.2 from *Reliability Engineering and Safety System*, **17**, 127–137, Ansell, R.O. and Ansell, J.I., Modelling the reliability of sodium sulphur cells, (1987), Table 6.1 from *Reliability Engineering and Safety System*, **21**, 293–308, Winfield, D.J., Long term reliability analysis of standby diesel generators, (1988), Table 10.4 from *Reliability Engineering and Safety System*, **14**, 63–74, Triner, D.A., The assessment of fleet equipment reliability, (1986), and material in Chapter 10 from *Reliability Engineering and Safety System*, **28**, 337–356, Ansell, J.I. and Phillips, M.J., Practical reliability data analysis, (1990). We acknowledge the permission of the Royal Statistical Society to reproduce material from *Applied Statistics*, **38**, 205–247, Ansell, J.I. and Phillips, M.J., Practical problems in the statistical analysis of reliability data, (1989). We would also like to make the following acknowledgments. Table 1.3 is reprinted from *Technometrics*, **32**, 67-82, Keating, J.P., Glaser, R.E. and Ketchum, N.S., Testing hypotheses about the shape parameter of a gamma distribution, (1990), with the permission of the American Statistical Association and the American Society for Quality Control. Tables 1.4 and 1.6 are reprinted from *IEEE Transactions on Reliability*, **40**, 134-139, Kim, J.S. and Proschan, F., Piecewise exponential estimator of the survivor function, (1991), and *IEEE Transactions on Reliability*, **39**, 329-335, Elsayed, E.A. and Chan, C.K., Estimation of thin-oxide reliability using proportional hazards models, (1990), respectively, with the permission of the Institute of Electrical and Electronic Engineers, Inc. Table 1.5 is reprinted from *Journal of Quality Technology*, **22**, 230-238, Nelson, W., Hazard plotting of left truncated life data, (1990), with the permission of the American Society for Quality Control. Table 8.2 is reprinted from the *Proceedings of the 10th Advances in Reliability Technology Symposium, Bradford* (ed. G.P. Libberton), 252-262, Hughes, R.P., A framework for dependent failure analysis, (1988), with the permission of Chapman and Hall Limited, London.

We are also grateful to Paul Burton for his detailed comments on the first three chapters. MJP would like to acknowledge the hospitality given to him by Peter Hall and all his colleagues at the Centre for Mathematics and its Applications, Australian National University, Canberra, during the second half of 1993 when this text was completed. We would also thank the staff of Oxford University Press for their patience and assistance in completing this text.

Jake Ansell
Michael J. Phillips
May 1994

Contents

1
Introduction to reliability

1.1 Introduction

The objective of this book is to describe elements of statistical data analysis which are useful in reliability studies, in a manner which gives the flavour of modern approaches. The chapter commences with a description of three examples of reliability studies taken from the authors' practical experience of statistical data analysis. These studies, which cover a range of industrial contexts, are used later in Chapter 10 to illustrate the concepts presented throughout the book. It is important to define the objectives of a study and the objectives that can be achieved through the application of probability and statistical methods are outlined. The different forms of the data encountered to achieve the objectives of a reliability study are illustrated in Section 1.4. These illustrations are made using examples from recently published papers in reliability journals and include right censored data, truncated data, accelerated failure data, and data from repairable systems. All these data must be collected and the proper planning of the way these are collected is discussed. This leads to the consideration of databases, both the use of historic databases and the construction of new ones. However, before commencing any discussion of statistical methods a formal definition of *reliability* is required. The *reliability* of a system (or component) is defined as the probability that the system operates (performs a function under stated conditions) for a stated period of time. Usually the period of time is the initial interval of length t, which is denoted by $[0, t)$. In this case the reliability is a function of t so that the *reliability function $R(t)$* can be defined as

$$R(t) = P(\text{System operates during } [0, t)),$$

where $P(A)$ denotes the probability of an event A, say. To understand this definition it is necessary to understand the concept of probability and simple probability calculations. The necessary probability results are summarized and the concept of a random variable is also introduced in Section 1.6. This enables the various features of continuous distributions, which are

1

Table 1.1. Days on which an event took place for the event process with covariates

30	31	34	47	55	62	64	66	68	101
102	103	104	105	108	112	115	117	118	122
124	125	126	127	129	130	131	132	135	136
137	139	141	142	143	150	152	157	165	166
167	170	171	173	180					

used to model failure times, to be introduced. The relationships between the reliability function, the probability density function, and the hazard function are detailed. These ideas are illustrated by the exponential distribution, which is the simplest example of a failure time distribution. The chapter is concluded with an account of statistical methods and how they may be used in achieving the objectives of a reliability study. This covers the use of parametric, semi-parametric, and non-parametric models, though the different philosophical ways of answering the inferential questions posed by the use of these models are left until Chapter 2.

1.2 Examples of reliability studies

When describing a set of techniques it is important to demonstrate the techniques through case studies to highlight the questions which surround the approach to data analysis as well as their application. Three reliability studies described by Ansell and Phillips (1989, 1990) are presented for illustration, though they do not cover all the features that reliability data from studies may possess. So other examples will be given in Section 1.4. As far as possible for each study the practical setting is described as well as the client's objective.

1.2.1 PROCESS EVENT STUDY

A company wished to be able to predict the rates of events which affected the reliability of its processes. The data collected consist of dates of specific events during a 6 month period and are presented in Table 1.1. An initial inspection of the data suggests that the events (failures) occurred in one of two periods: firstly from 30 to 70 days and secondly from 100 to 180 days. The rate of occurrence is assumed to be related to two variables (covariates), either directly measured or derivable from the data collected. The objectives of the study are to ascertain whether the rate of occurrence

Table 1.2. Method of data collection for the repairable electronic systems

Collector	Data	Statistics
Manufacturer	Failures	EVENTS
	BAD NEWS	
Customer	Failure FREE time	CENSORED TIMES
	GOOD NEWS	

The systems were progressively introduced into service and NOT operated continuously or in a uniform manner.

of failures (events) was dependent on these variables and to be able to predict this rate.

1.2.2 ELECTRONIC SYSTEM STUDY

A manufacturer of electronic equipment decided to assess the performance of the systems supplied to a customer. The supplier's data in this case consisted of the failure times. The customer's data consisted of the failure free times. This is because whilst the manufacturer is under an obligation to repair failed systems for a customer and will therefore have information about these failed systems, the customer is under no obligation to supply information about systems which are performing satisfactorily. This information must be obtained from the customer in some direct or indirect way. The information from both sources, the BAD NEWS of the failures and the GOOD NEWS of the failure free times, was combined to produce the data. The method of data collection is summarized in Table 1.2. Another aspect of the data is that calendar time is not an appropriate time scale since the system is not used throughout the period. The time taken to repair the system is not taken into consideration as these times are short compared with the average time between failures.

When a failure of the system occurs in service the electronic module (subsystem) which caused the failure is identified and is then replaced by a new module and the system is returned to service. So there is a record

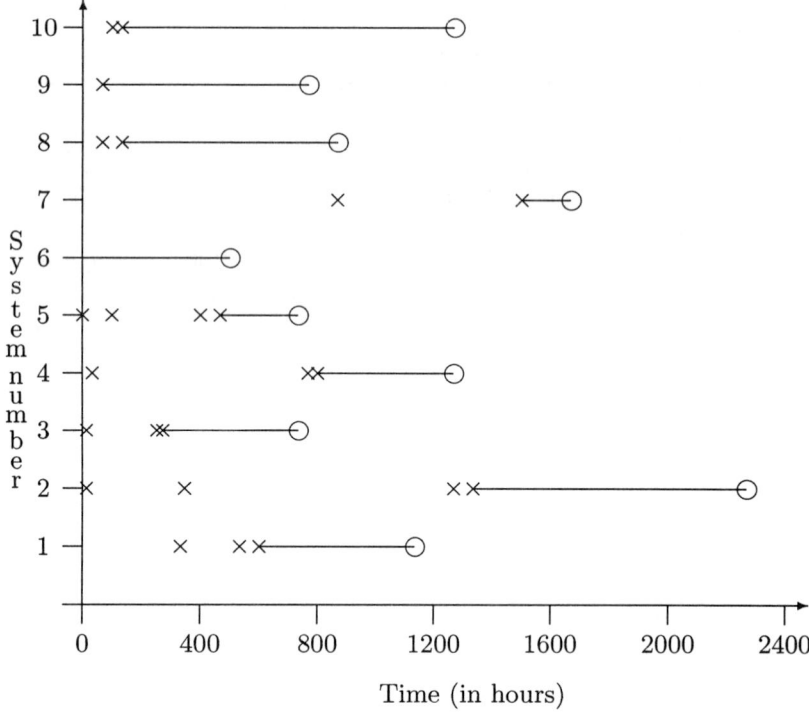

Fig. 1.1. Example of the data from a group of 10 of the repairable electronic systems. × : failure; ◯ : last time withdrawn; _____ : failure free time

for each system which consists of the failure times for each failure with the appropriate serial number of the replaced module. The record also contains an estimate of the last time at which the system was withdrawn from service before the end of the period of data collection. An example of a subset of these data is presented in Fig. 1.1, though the serial numbers of the replaced modules are not included. An inspection of the data shows that it is possible for failures to occur 'early' in the first 100 hours, though subsequently there are quite large failure free times, often of as much as 400 hours. The objectives of the study are to use these data on failure times and failure free times to decide whether the systems satisfy specification, need a major modification, or a complete redesign.

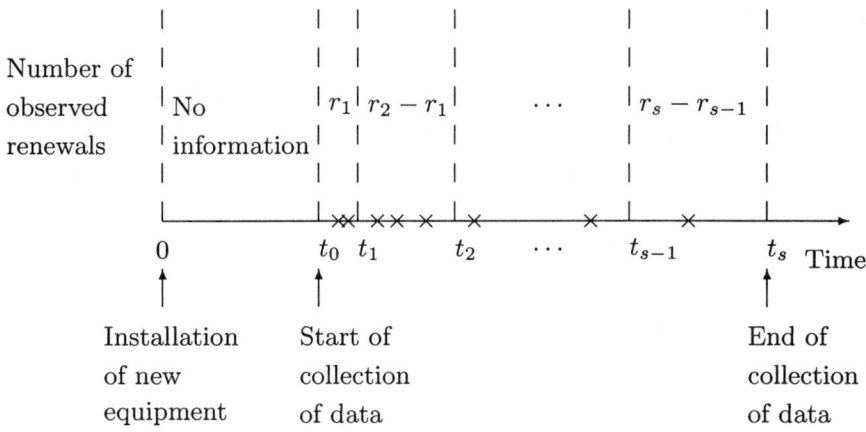

Fig. 1.2. Method of data collection for the mechanical equipment fitted to a fleet. × denotes an observed renewal

1.2.3 FLEET MECHANICAL EQUIPMENT STUDY

The operator of a fleet decided to assess the performance of mechanical equipment fitted to the ships from data which have been collected from the operation of the fleet over a number of years. The data consist of the number of events (failures) r_i in the ith $(1 \leq i \leq s)$ period (t_0, t_i), assuming instantaneous repairs. This is because the information was collected manually and was not sufficiently accurate to be more precisely recorded. Hence the number of events in the period (t_{i-1}, t_i) is $r_i - r_{i-1}$, with $r_0 = 0$. A feature of the data collection is that the number of failures in the initial period $(0, t_0)$ are not known as a decision to collect information was only taken at some time after the original fitting of the new equipment. The method of data collection is shown in Fig. 1.2 and the data are presented in Chapter 10 in Table 10.4. The problem was described by Triner (1986) as it contains features which are quite common to many sets of data collected on marine equipment. The data exhibit an increasing rate of failure as time increases as would be expected with mechanical equipment which is subjected to 'wear'. The objectives of the study are to decide about the future replacement or repair of the fleet equipment.

1.3   Objectives of reliability studies

In any reliability study there is a need to define the objectives or aims. The objectives will depend on several factors, such as who has instigated the study and why they have done so. Examples of possible objectives are:

(i) ascertaining whether a system has achieved a specific level of performance;

(ii) estimation of the cost of achieving a given level of performance of the system;

(iii) assessment of the likelihood of achieving a successful mission; or

(iv) decisions on the direction of resources to achieve the optimum results.

It may be that the objectives are a combination of some of the above or of other objectives.

It is not surprising that many objectives in a reliability study may be achieved without recourse to statistical analysis. However, this book will concentrate on objectives which can be achieved through the application of probability and statistical methods. The three main concerns in statistical data analysis in reliability studies are assessment, identification, and prediction. Assessment is concerned with the estimation of the distribution of the component or system lifetimes, or the probability of successful completion of a task. Identification is the process of identifying the significant features which affect lifetimes, or of the components which require improvement to enhance performance. Prediction is the extrapolation into the future based on historical data. One or more of these concerns will be appropriate depending on the specified objectives of the study.

The objective of the study should always be kept in mind when carrying out the statistical analysis. The approach should always be 'problem-led' rather than 'technique-led'. So there is no need for over-elaboration in employing techniques as frequently simple techniques can achieve the desired objective. Rarely will a manager or engineer need to know the specific form of a lifetime distribution though they will be interested in whether reliability has improved. However, this does not mean that the statistical methods required to achieve an objective are always simple!

1.4   Nature of reliability data

To achieve the objectives of a reliability study it is necessary to obtain facts from which the other facts required for the objectives may be inferred. These facts or *data* obtained for reliability studies may come from testing under factory/laboratory conditions or from field studies. A num-

Table 1.3. Times to failure (in hours) for pressure vessels

274	28.5	1.7	20.8	871	363	1311	1661	236	828
458	290	54.9	175	1787	970	0.75	1278	776	126

ber of examples of different kinds of reliability data obtained in reliability studies will now be given. Rather than use the data of the examples of reliability studies given in Section 1.2, which will be discussed in detail in Chapter 10, examples will be given which are taken from recently published papers in the following well-known reliability journals: *IEEE Transactions in Reliability, Journal of Quality Technology, Reliability Engineering and Safety Systems*, and *Technometrics*. These journals are readily accessible enabling the reader to consult them for further details and analyses. It is unfortunately the case for the development of the application of statistics to reliability that commercial sensitivity is often a cause of lack of published data. This is being corrected, but still many novel and useful techniques are not explored through lack of data.

1.4.1 EXAMPLE OF SINGLE SAMPLE FAILURE DATA

The simplest example of failure time data is a single sample taken from a number of similar copies of a system. An example of this was given by Keating *et al.* (1990) for 20 pressure vessels, constructed of fibre/epoxy composite materials wrapped around metal liners. The failure times (in hours) are given in Table 1.3 for 20 similarly constructed vessels subjected to a certain constant pressure.

This is an example of a complete sample obtained from a factory/laboratory test which was carried out on 20 pressure vessels and continued until failure is observed on all vessels. There is no special order for the failure times. These data were used by Keating *et al.* (1990) to show that the failure distribution could have a decreasing hazard function and that the average residual life for a vessel having survived a wear-in period exceeds the expected life of a new vessel.

1.4.2 EXAMPLE OF RIGHT CENSORED FAILURE DATA

If one of the objectives of collecting data is to observe the time of failure of a system it may not always be possible to achieve this owing to a limit on

Table 1.4. Times to failure and censored times (in days) for telecommunication systems during five months in 1985

164+	2	45	147+	139+	135+	3	155+	150+	101+
139+	135+	164+	155+	150+	146+	139+	135+	164+	155+
150+	1	139+	7	163+	139	149+	143+	138+	134+
163+	152+	149+	143+	40	13	163+	152+	149+	143+
138+	134+	163+	152+	149+	142+	138+	134+	163+	152+
149+	10	138+	134+	163+	94	149+	141+	138+	133+
77	151+	149+	141+	138+	133+	162+	151+	149+	141+
138+	133+	162+	151+	149+	34	138+	133+	73+	151+
115	140+	138+	133+	63	151+	138+	140+	137+	133+
161+	151+	148+	140+	137+	64+	160+	151+	147+	140+
137+	133+	160+	151+	147+	140+	137+	133+	67	90+
147+	140+	137+	133+	141+	151+	147+	140+	137+	133+
156+	151+	147+	54	137+					

Note: Figures with + are right censored observations, not failures.

the time or resources spent on the data collection process. This will result in incomplete data as only partial information will be collected, which will occur if a system is observed for a period and then observation ceases at time C, say. Though the time of failure T is not observed it is known that T must exceed C. These forms of incomplete data are known as *right censored* data (see Chapter 2, Subsection 2.4.1).

An example of these kinds of data was given by Kim and Proschan (1991) for telecommunication systems installed for 125 customers by a telephone operating company. The failure times (in days) are given in Table 1.4 for 16 telecommunication systems installed in 1985 and the censored times (when the system was withdrawn from service without experiencing a failure or the closing date of the observation period) for the remaining 109 systems. The systems could be withdrawn from service during the observation period on the customer's request. As well as failure to function, an unacceptable level of static, interference, or noise in the transmission of the telecommunication system was considered as a system failure.

This situation where a high proportion of the observations are censored is often met in practice. These data were used by Kim and Proschan (1991) to obtain a smooth non-parametric estimate of the reliability (survivor) function for the systems (see Chapter 2, Section 2.5).

Table 1.5. Times to first failure (in years) for compressors in heat pumps between 2.59 years and 9.33 years

---

3.30 4.62 4.62 5.75 5.75 7.42 7.42 8.77 9.27 9.27

The remaining 154 compressors have right censored observations greater than 9.33.

1.4.3 EXAMPLE OF TRUNCATED FAILURE DATA

If the time of failure T is not observed if it occurred in some interval of time then these forms of data are known as *truncated* data (see Chapter 2, Subsection 2.4.2). An example of these kinds of data was given by Nelson (1990) for compressors in heat pumps which were installed in a number of buildings under service contract. Compressor data were collected only while a building was on service contract and if this only started after the compressors had been in service for a time then there is a 'gap' in the data. The data on compressors in such a building have a common age when they start being observed and are said to be *left truncated*. The failure times (in years) of first failure are given in Table 1.5 for 10 compressors which failed after 2.59 years and the remaining 154 compressors failed after 9.33 years and had been installed in the same building.

If data were obtained from a number of buildings then the pooled data will be multiply left truncated and right censored. Nelson (1990) used these data to construct hazard plots, which can be used for a non-parametric estimate of the reliability (survivor) function. The objectives of the study were to estimate the median life of the compressors and to assess whether the hazard function increases or decreases with time.

1.4.4 EXAMPLE OF ACCELERATED FAILURE DATA WITH COVARIATES

Elsayed and Chan (1990) presented failure times for the time-dependent dielectric breakdown of metal–oxide–semiconductor integrated circuits for accelerated failure testing. The objective was to estimate the dielectric reliability and hazard functions at operating conditions from the data obtained from the tests. Tests were performed to induce failures at three different elevated temperatures where temperature is used as the stress variable (or

Table 1.6.   Times to failure and censored times (in hours) for three different temperatures

Temperature (170 °C)									
0.2	5.0	27.0	51.0	103.5	192.0	192.0	429.0	429.0	954.0
2495.0	2495.0	2495.0	2495.0	2495.0	2495.0	2495.0	2495.0	2948.0	2948.0+

Temperature (200 °C)									
0.1	0.2	0.2	2.0	5.0	5.0	25.0	52.5	52.5	52.5
52.5	123.5	123.5	123.5	219.0	219.0	524.0	524.0	1145.0	1145.0

Temperature (250 °C)									
0.1	0.2	0.2	0.2	0.5	2.0	5.0	5.0	5.0	10.0
10.0	23.0	23.0	23.0	23.0	23.0	50.0	50.0	120.0	

Note: Figure with + is a right censored observation, not a failure.

covariate). These data for silicon dioxide breakdown can be used to investigate failure time models which have parameters dependent on the value of the covariate (temperature) and are given in Table 1.6. The nominal electric field was the same for all three temperatures. Elsayed and Chan (1990) used these data to estimate the parameters in a semi-parametric proportional hazards model (see Chapter 3, Section 3.5).

1.4.5   EXAMPLE OF REPAIRABLE SYSTEM FAILURE DATA

The examples in Subsections 1.4.1 to 1.4.4 of failure data are records of the time to first failure of a system. Some of these systems, such as the pressure vessels described in Subsection 1.4.1 and the integrated circuits described in Subsection 1.4.4, may be irreparably damaged though others, such as the telecommunication systems described in Subsection 1.4.2 and the compressors described in Subsection 1.4.3, will be repaired after failure and returned to service. This repair is often achieved by replacing part of the system to produce a *repairable* system (see Chapters 5 and 6).

Table 1.7.   System failure times (in hours)

System A -	452	752	967	1256	1432	1999	2383	(3000)
System B -	233	302	510	911	1717	2107		(2500)
System C -	783	1805						(2000)
System D -	782							(1500)
System E -								(1000)

Note: Figures in parentheses are current cumulative times.

Newton (1991) presented failure times for a repairable system. Five copies of a system were put into operation at 500 hour intervals. When failures occurred the component which failed was instantaneously replaced by a new component. Failures were logged at cumulative system hours and are given in Table 1.7. The times of interest are the times between the times of system failure.

The objective for which Newton (1991) used these data was to consider the question of whether the failure rate for the systems was increasing or decreasing with time as the systems were repaired.

1.5   Databases and data collection

The collection of data needs to be planned properly before a study if the objectives of the study are to be successfully achieved. A central theme throughout the development of approaches to reliability analysis is the need for *reliable* information. The construction of *databases* has been seen by many as the solution to this problem and a number have therefore been constructed. However, these databases would have been constructed with the technology available at the time of their implementation, and they will have been designed to the state-of-the-art requirement. Generally, though there are exceptions, these databases have not withstood the test of time. Many of the original databases would have been hierarchical in structure allowing only limited enquiries of the data to be made. Designers of modern databases have the advantage of sophisticated software and a better comprehension of the design requirement for a reliability database.

1.5.1   HISTORIC DATABASES

Historic databases have suffered from a number of drawbacks. These will now be discussed under six headings.

(i) *The use of secondary data*
Most reliability field data are derived from alert/incident reports or maintenance/inventory records. Hence the main concern of the collector of the information is not the production of a reliability database but either the analysis of the alert or reporting the maintenance action. This means that the data are derived, or secondary. Hence the data collector will not necessarily take into account the requirements of the reliability analysts.

Whilst alert/incident reports are useful qualitatively to indicate the nature of alerts/incidents they do not necessarily give the frequency of the contributing events. This is because only the events which give rise to alerts are recorded. Also such a report does not assist in assessing the frequency of potential alerts/incidents, as it gives only those that have been noted. The information available will have been collected during or subsequent to an alert/incident and this will affect the quality of the data obtained and also the material which is recorded.

Maintenance (inventory) reports are less misleading in that they do not focus on a single point in time such as an alert. Hence it is more reasonable to use them to assess the frequency of occurrence. However, the information which is often recorded is limited purely to the concerns of the maintenance engineer. A typical record is presented in Table 1.8, which highlights another problem, namely that the reports suffer from subjective assessments of faults and relationships. Failures are noted when they have been reported for maintenance and the time noted with an event is often only the date of the start of work and the end of work. Rarely is the time that the failure was noted or revealed recorded. Also the example in Table 1.8 illustrates that a maintenance operator may be unable to diagnose a fault or to report it within the categories required by the data collection system. Hence under the heading CAUSE there are the entries UNKNOWN and SEE NOTE. This latter comment, as Bendell and Walls (1985) have pointed out, may be very significant in revealing the failure mechanism! This information is often lost when transferred to databases.

(ii) *Generic data collection*
Some databases were constructed so that the information would be either stored or only capable of being processed for a generic type of component/system. Hence the context of the data was ignored. Information therefore on components within benign contexts would be combined with information on the same type of component within a severe context. For example, data on compressors might arise from operation on both land and

Table 1.8. A typical maintenance record

Instrument no.	Date of event	Symptom	Cause	Failure modes	Item code	UNIT	How affected	Action taken	Work order no.
GA8000a/1	01 08 72	MAL-FUNCT	SEE NOTE	* S R	MI	TEST GEAR	LOOSE	TIGHT-END	72000560
GA8000b/2	25 12 74	TRIP-FLT	UN-KNOWN	B S R	**	INDIC-ATR	SUSPECT	REPAIR-ED	74000978
GA8000A/1	05 04 76	PROC-LEAK	PROC-RATE	R S U	WI	DP CELL	LEAKING	NONE	76000146

sea. Data used for reliability analysis must come from a context which is relevant to the study being undertaken.

(iii) *The use of qualitative data*
One reason for the early databases' inability to disaggregate data was the lack of the capability to analyse qualitative data. Whilst it was possible to store qualitative data, and many took advantage of such facilities, it was not necessarily appreciated that such data could be analysed or were even worthy of analysis. As has been shown already the qualitative data could have a major impact on the analysis and understanding of the data. The loss of the context and the details of the special cases invalidate the use of some databases.

(iv) *Clarification of definitions*
Another problem is the lack of clarity contained within the definitions used to describe the data. Modes of failure, symptoms, and causes are often ill defined and ambiguous. Sometimes components/systems are not clearly specified. There have been examples (Newton, 1991) where the serial numbers of systems have been 'randomly' assigned after overhauls so that the analysis of data based on these serial numbers would be very misleading!

(v) *Validation of the data*
Some of the early databases did not check or validate the data before they were added to the database so that *negative* lifetimes could be obtained! These would obviously be picked out when any analysis was performed, though other incorrect but apparently sensible values would not.

(vi) *Relevance of the data*
Before using a database it is necessary to ensure that it is relevant to the study being undertaken. Firstly it needs to be established if technological or procedural improvements to the component/system have taken place since the database was established. Secondly are the data collected relevant to future studies? The second point is more subtle. O'Connor (1991)

quotes the example of microchips and suggests that in some cases the data collected refer to a subset of the components which would be rarely used in practice. Early failures of microchips might be associated with 'faulty' chips and such chips might be eliminated by inspection and hence never be used. The lifetimes of chips used would then be misrepresented if these early failures were included. So it is important to stress the need for care in using historic databases. If the data are considered relevant and appropriate then it would be wise to validate them before using them in an analysis.

1.5.2   CONSTRUCTION OF A DATABASE

For the reasons that have been detailed it may be sensible to construct a database for a study from known sources. The following points should be borne in mind if this is to be done. These points are based on the comments of Moss (1991) and Bendell and Walls (1985).

In order to ensure a 'good' source of data it is important that a data quality assurance plan be adopted. This would require a structured approach in a number of specific areas: organization, documentation, data collection, data processing, and reporting. As with the objective of the reliability study it is necessary to be clear about the objective of establishing the database. Because of future demands on the database it may be unwise to restrict data collection unduly to the specific study in progress.

It is necessary not only to clarify the objective of the study but also to decide how it is to be achieved. The roles of those involved with the collection should be established and where appropriate necessary training should be implemented. The realization by those involved with the exercise of the importance of the task and the objectives will hopefully motivate the individuals involved. It may well be that those directly involved with collection will be more aware of practical issues and this may be crucial to the exercise.

Documentation should be viewed as assisting with the data collection rather than as a bureaucratic exercise. Documentation should remove ambiguity, clarify the objectives, amplify the requirements of the process, and should contain a full list of the components and systems on which data are to be collected. The database should be described in detail, including the entry screens used for computer data entry, to ensure that there are no obvious ambiguities and all aspects involved in collection have been considered. The data which should be collected depend on the context and objective of the study. Some general areas are given in Table 1.9 as a guide. It is necessary to ensure that the full data requirements are identified and can where necessary be obtained. This will require a clear definition of the measurement to be made and how the data are to be collected and verified.

Table 1.9. Information requirements for reliability studies

Data requirements	Further details
Component / system data	
Equipment identification	Type / serial number
Boundary specification	Location limits
Design details	
Operation parameters	Installation / failures / repairs
Description definition	
Failure data	
Time	Calendar / test time / operating time
Number of demands	Starts / stop / tests
System operating states	Operable / repair
State / time correspondence	
Time on standby	
Failure mode	
Symptom	
Cause	
Special operational / environmental conditions	
Diagnostic covariates	
Maintainability data	
Operatives	
Active repair times	
Waiting times	
Special tools / equipment	
Types of maintenance	

For example, it is possible to have automatic and manual checks on times. Where checking is appropriate it is always sensible to ensure that there is direct feedback to those involved in the collection process. This will assist in improving the quality of the data and highlight problems which require tackling.

Before data are accepted they should be validated and checked for consistency. Validation will depend on a clear definition of unacceptable data. Checks that ensure no inconsistencies within and between samples are valuable. Also validation using other databases, where possible, can be productive. Elementary data analysis, when included in the data processing, is sensible as it often will show up problems with the data collection. Only when data have passed through positive vetting is it wise to add them to the database. Then more detailed analyses may be carried out on the accepted data.

To ensure that the data collection process is being carried out as expected it is important to have a reporting process. This should involve regular progress reports on a suitable time scale. Too frequent reporting can give rise to problems as can no reporting. For any data collection exercise a final or main review stage should be planned at the beginning of the exercise. It should be clearly laid out what form is required for the data, including the details of the data summaries.

1.6   Probability and random variables

Reliability has been defined as a probability, which is a function of time, used for the analysis of the problems encountered when studying the occurrence of events in time. Several aspects of time may be important: (i) age, (ii) calendar time, (iii) time since repair. However, for simplicity, problems are often specified in terms of one time scale, values of which will be denoted by t.

It is simplest to introduce the basic ideas of probability in the discrete case before considering probability as a function of continuous time.

1.6.1   PROBABILITY

Probability is one of the fundamental ideas used in *statistics*, the science of methods for data analysis, and had its formal beginnings in the study of games of chance in the 17th century. It is usually now introduced through the axiomatic approach due to Kolmogorov, which was excellently described by Feller (1968). This approach considers an *experiment* and every possible *outcome* (result) of the experiment. The set which is the totality of all possible *outcomes* will be defined as the *sample space*. An *event* is a collection of outcomes which can be considered as a subset of the sample space. To understand the relations between two or more events an understanding of some simple results of set theory is an advantage.

The following definitions will now be given for events (or sets as events are defined as sets).

Definitions of events (sets)
(i) The *complement* of the event A, which is the event that A does **not** occur and so is defined by the outcomes which are **not** contained in A, is denoted by $\bar{A}$.

(ii) The *union* of two events A and B, which is the event that A or B **or** both A and B occur and so is defined by the outcomes which are contained in A or B **or** both A and B, is denoted by $A \cup B$.

(iii) The *intersection* of two events A and B, which is the event that **both** A and B occur, is denoted by AB.

Probability is defined by means of the following axiom.

The axiom of probability

For every outcome E_i, $i = 1, 2, ...$, in the countable sample space a real non-negative number $P(E_i)$, the probability of E_i, is defined. These probabilities are assigned so that their sum, $\sum_i P(E_i)$, is unity.

Hence every outcome has a probability which is not greater than one. This definition can be extended to events by defining the probability $P(A)$ of an event A, say, to be equal to the sum of the probabilities of the collection of outcomes which define A.

Often, having defined probability with respect to one sample space, it is desired to restrict attention to outcomes from a subset of the sample space (corresponding to the event B, say). It would be possible to redefine the probabilities with respect to this new sample space. However, it is simpler to use *conditional* probability.

Definition of conditional probability

Provided $P(B)$, the probability of the event B, is positive the *conditional* probability of the event A given B is defined by

$$P(A \mid B) = P(AB)/P(B),$$

where $P(AB)$ is the probability of the event given by the *intersection* of A and B, which is defined by the outcomes which are contained in **both** the events A and B.

It is possible that the redefinition in terms of B has no effect on $P(A)$. This leads to the idea of the *statistical independence* of events. Two events are *statistically independent* if the product rule

$$P(AB) = P(A)P(B)$$

applies, and hence $P(A \mid B) = P(A)$.

1.6.2 RANDOM VARIABLES

Probability has been defined for events using the outcomes of an experiment. Often the value of a variable X, say, is associated with every outcome. Then if the only interest is in events which can be defined in terms of values of X then it is only necessary to know the values of the probability associated with every value x of the variable X, $P(X = x)$ say. This leads to the concept of a *random variable* X, which is a function defined on a sample space, and its probability distribution which for a discrete sample space is defined by a *probability function*. (The values of x will be assumed to be integer valued.)

Definition of the distribution of a discrete random variable
A discrete random variable X has a distribution with a *probability function* $p_X(x)$, for $x = 0, 1, 2, \ldots$, where

$$p_X(x) = P(X = x).$$

Example 1.6.1
An example where the binomial distribution is used is when systems are classified into two types: failed and unfailed systems. If there are n systems and the probability that a system fails is p (and hence the probability that a system does *not* fail is $1 - p$) then X, the number of systems that fail, has a binomial distribution. This means that the probability function is given by

$$p_X(x) = \binom{n}{x} p^x (1 - p)^{n-x}, \qquad \text{for } 0 \leq x \leq n,$$

where

$$\binom{n}{x} = \frac{n!}{(n - x)!x!},$$

and $x!$ is the factorial of x given by $x! = x(x - 1) \ldots 2.1$.

□

Example 1.6.2
If in Example 1.6.1 the number of systems, n, increases indefinitely to infinity in such a way that the probability of failure, p, decreases but the product np remains constant and equal to μ, say, then Y, the number of systems that fail, has a Poisson distribution. So the probability function is given by

$$p_Y(y) = \frac{\mu^y \exp(-\mu)}{y!}, \qquad \text{for } y = 0, 1, 2, \ldots.$$

In principle the value of Y can take any non-negative integer value.

□

Often it is necessary to obtain the probability of an event which is defined by restricting values of the random variable X to an interval $x =$

$x_1, x_1 + 1, x_1 + 2, ..., x_2$, say. This can be obtained by summation of the probability function so that

$$P(x_1 \leq X \leq x_2) = \sum_{i=x_1}^{x_2} p_X(i).$$

So it is possible to obtain the probability of an event defined by the random variable X by suitable summation of the probability function.

1.6.3   THE DISTRIBUTION OF THE FAILURE TIME

In order to study the reliability of a component, define a random variable T denoting the time of occurrence of an event of interest (failure). Then the event 'system operates during $[0,t)$' used in Section 1.1 to define the reliability function is equivalent to '$T \geq t$'. Then the reliability function of the failure time distribution can be defined by

$$R_T(t) = P(T \geq t),$$

and is a non-increasing function of t. The reliability of a component cannot improve with time!

The reliability function can be used to obtain the probability that the failure occurs in an interval of time $[t_1, t_2)$, say, as

$$P(t_1 \leq T < t_2) = P(T \geq t_1) - P(T \geq t_2) = R_T(t_1) - R_T(t_2).$$

For a continuous random variable like T the reliability function should be a 'smooth' continuous function, if there are no time points at which a failure is likely to occur exactly at that precise time, and this implies that the probability of the event '$T = t$' is always zero! This may at first seem a paradox, but it is a consequence of the continuity of the reliability function as can be seen by using the result for $P(t_1 \leq T < t_2)$ with $t_1 = t$ and $t_2 = t + h$ and considering the limit as h tends to zero. (However, this is not a *practical* problem as any failure time can only be measured to the smallest unit of measurement of the 'chronometer' and therefore must be considered as belonging to an interval of time, however small.) Hence the distribution of T cannot be defined using a probability function as was done for discrete distributions in Subsection 1.6.2. So instead of the probability function the *probability density function* of the failure time distribution is used and this is defined by

$$f_T(t) = \lim_{h \to 0} \frac{P(t \leq T < t + h)}{h} = \frac{-dR_T(t)}{dt}.$$

This function is **not** a probability and can take values greater than unity, though it can be manipulated like a probability because it can be used as an approximation, as

$$P(t \leq T < t + h) \approx f_T(t)h,$$

for small h. So when multiplied by h the probability density function gives (approximately) the probability that T, the time of failure, occurs 'just' after time t (in $[t, t+h)$).

The probability density function is obtained from the reliability function by using *differential calculus*. This process can be reversed so that by using *integral calculus* the reliability function can be obtained from the probability density function. Therefore

$$
\begin{aligned}
P(t_1 \leq T < t_2) &= R_T(t_1) - R_T(t_2) \\
&= \int_{t_1}^{t_2} f_T(u)du.
\end{aligned}
$$

This can be compared with the result obtained for the probability of an event defined by the interval of a discrete random variable which was given in Subsection 1.6.2 by a summation. In general for a continuous random variable summation is replaced by integration (as this is the limiting case of summation as the discrete interval between values of the discrete variable tends to zero).

Another function which can be used to define the distribution of T is the *hazard function* (*age-specific failure rate* or *force of mortality*) of time to failure, which is defined by

$$\lambda_T(t) = \lim_{h \to 0} \frac{P(t \leq T < t + h \mid T \geq t)}{h} = \frac{f_T(t)}{R_T(t)}.$$

Again this function can be used to approximate a probability, in fact the conditional probability

$$P(t \leq T < t + h \mid T \geq t) \approx \lambda_T(t)h,$$

for small h. So when multiplied by h the hazard function gives (approximately) the probability that T, the time of failure, occurs 'just' after time t (in $[t, t+h)$) **given** it had not occurred up to time t. The hazard is also known as the *risk of failure*.

A related function is the *cumulative (integrated) hazard function* of the failure time distribution, which is closely connected with the reliability function and is defined by

$$\Lambda_T(t) = \int_0^t \lambda_T(u)du = -\log(R_T(t)).$$

(Note: log will be used to denote the natural logarithm to the base e $\approx$ 2.718.) So the reliability function can be obtained from the cumulative hazard function by

$$R_T(t) = \exp(-\Lambda_T(t)).$$

Example 1.6.3

One of the simplest examples of a distribution for a *positive* continuous random variable is the *exponential* distribution. This distribution has a reliability function which is given by

$$R_T(t) = \begin{cases} \exp(-\lambda t), & \text{for } t \geq 0, \\ 1, & \text{for } t < 0, \end{cases}$$

for $\lambda > 0$. By differentiating the reliability function with respect to t the probability density function is given by

$$f_T(t) = \begin{cases} \lambda \exp(-\lambda t), & \text{for } t \geq 0, \\ 0, & \text{for } t < 0. \end{cases}$$

Hence λ is the value of the hazard function which is therefore **constant** for all $t \geq 0$. This implies that the distribution possesses the famous *'lack of memory'* property as the risk of failure, which is the probability of a failure occurring 'just' after time t (given it has not yet occurred up to time t), remains the same for every time t. This property is special to this distribution and hence *characterizes* the distribution, i.e. it is not possessed by any other continuous distribution. The constant hazard function implies that the cumulative hazard function is a *linear* function of t, so that

$$\Lambda_T(t) = \lambda t, \text{ for } t \geq 0.$$

There is a close connection between this distribution and the Poisson distribution given in Example 1.6.2. The exponential distribution is the distribution of the times between events which occur 'randomly' in time as a Poisson process, which is a stochastic (random) process where the number of events in a fixed interval of time of length t has the Poisson distribution given in Example 1.6.2 with $\mu = \lambda t$.

□

Further examples of continuous distributions are the gamma, Weibull, extreme value, and normal (Gaussian) distributions which are discussed in detail in Chapter 2, Section 2.3.

These results which have been presented are for the *univariate* case of **one** random variable T. When a number of observations $T_1, T_2, ..., T_n$ of a random variable T are obtained then the *joint* (or *multivariate*) distribution of these n random variables will be required. If these observations have been obtained *independently* then events defined by the different T_i will also be *independent*. Then the observations are known as a *random sample*. As was seen at the end of Subsection 1.6.1 the probabilities of independent events satisfy a product rule. Hence this rule will be true for the reliability and probability density functions of the joint distribution. If the n random variables have the **same** distribution, as T say, then for the joint distribution of $T_1, T_2, ..., T_n$ the reliability function is

$$R_{T_1, T_2, ..., T_n}(t_1, t_2, ..., t_n) = \prod_{i=1}^{n} R_T(t_i)$$

and the probability density function is

$$f_{T_1, T_2, ..., T_n}(t_1, t_2, ..., t_n) = \prod_{i=1}^{n} f_T(t_i).$$

So in this *independent* case these two multivariate functions are obtained by using the *product* of the univariate functions.

These results will be used when likelihood methods of estimation are discussed in Chapter 2, Subsection 2.6.3.

1.7   Statistical methods

To make estimates or inferences based on a parametric model *information* is collected in the form of reliability data. The most effective method of achieving this is by collecting a number, n say, of observations, known as a *sample*, which are *independent*. The advantage of independent observations is that completely new information is obtained from each new observation, which would not be the case if the observations were dependent.

Definition of a random sample
The collection of *independent* observed values $t_1, t_2, ..., t_n$ of a random variable T, say, is known as a *random sample*. The joint distribution of the n observations has a reliability function and probability density function which are given by the *product* rules given at the end of Subsection 1.6.3.

A function of the observations from a sample is known as a *statistic*. Statistics can be used to summarize the information from the sample and the aim of *statistical methods* is to give ways in which these statistics may be used in answering questions arising from estimation and inference.

Example 1.7.1
Two of the most frequently used statistics are the *sample mean* $\bar{t}$ given by

$$\bar{t} = \frac{\sum_{i=1}^{n} t_i}{n}$$

and the *sample variance* s^2 given by

$$s^2 = \frac{\sum_{i=1}^{n}(t_i - \bar{t})^2}{n-1}.$$

The sample standard deviation is the positive square root of the sample variance.

□

1.7.1 PARAMETRIC MODELS

When considering possible failure time distributions to model the behaviour of observed failure times, it is not usual to choose a fully specified function to give the reliability or probability density function of the distribution but instead to use a class or 'family' of functions. This family of functions will be a function of some variables known as *parameters*, whose values have to be specified to 'index' the particular function of the family to be used to define the failure time distribution.

Example 1.7.2
For the two discrete distributions given as examples in Subsection 1.6.2 the parameters for the binomial distribution in Example 1.6.1 are n and p, and for the Poisson distribution in Example 1.6.2 the parameter is μ.

For the continuous failure time distribution in Example 1.6.3 the parameter of the exponential distribution is λ, the hazard function.

Examples of other 'families' of distributions are the exponential, Weibull, and normal (Gaussian) distributions which are discussed in Chapter 2, Section 2.3.

□

More than one parameter may be used and then the r parameters, say, will in general be denoted by the vector $\boldsymbol{\beta} = (\beta_1, \beta_2, ..., \beta_r)^T$ and the reliability or probability density function of the parametric distribution for a continuous failure time will be denoted by $R_T(t; \boldsymbol{\beta})$ and $f_T(t; \boldsymbol{\beta})$, respectively.

One of the objectives of a reliability study may be to study the behaviour of the failure time distribution of T through the variation of k other variables represented by the vector $\mathbf{z} = (z_1, z_2, z_3, ..., z_k)^T$, say. These variables are usually referred to as *covariates*. In order to perform any parametric statistical analysis it is necessary to know the joint distribution of the failure time T and the covariates $\mathbf{Z}$ whose probability density function will be denoted by $f_{T,\mathbf{Z}}(t, \mathbf{z}; \boldsymbol{\beta})$, for parameters $\boldsymbol{\beta}$.

For any system it may be possible to obtain the covariate values $\mathbf{z}$ and then the interest is focused on the failure time distribution **given** these values, which is the *conditional* distribution of T given $\mathbf{z}$. In an analogous way to the definition of conditional probability for two events presented in Subsection 1.6.1, the probability density function of the conditional distribution of T given $\mathbf{z}$, $f_{T|\mathbf{Z}}(t \mid \mathbf{z}; \beta)$, is given as the ratio of two probability density functions because it was seen in Subsection 1.6.3 that probability density functions could be manipulated in the same way as probabilities. Hence

$$f_{T|\mathbf{Z}}(t \mid \mathbf{z}; \beta) = \frac{f_{T,\mathbf{Z}}(t, \mathbf{z}; \beta)}{f_{\mathbf{Z}}(\mathbf{z}; \beta)},$$

where

$$f_{\mathbf{Z}}(\mathbf{z}; \beta) = \int_0^\infty f_{T,\mathbf{Z}}(t, \mathbf{z}; \beta) dt.$$

So in the denominator $f_{\mathbf{Z}}(\mathbf{z}; \beta)$, the probability density function of the distribution of $\mathbf{Z}$ is obtained from the $f_{T,\mathbf{Z}}(t, \mathbf{z}; \beta)$, the probability density function of the joint distribution of T and Z, by integrating out t.

In general, the probability density function of any subset of a set of random variables can be obtained from the probability density function of the full set by integrating out those random variables not required. The resulting distribution is commonly referred to as the 'marginal' distribution.

Definition of a failure time model

A failure time model specifies the distribution of the failure time T conditional upon the covariates $\mathbf{z}$ using either the conditional probability density function $f_{T|\mathbf{Z}}(t \mid \mathbf{z}; \beta)$ or the conditional reliability function $R_{T|\mathbf{Z}}(t \mid \mathbf{z}; \beta)$, with parameters β.

Often it is necessary to relate failure to variables which vary with time. These are called *time-dependent covariates*. The *covariate history* at time t consists of the values of the covariate $\mathbf{z}$ up to time t and is denoted by $\mathbf{z}(t)$.

Definition of a hazard model

A hazard model uses $\lambda_{T|\mathbf{Z}}(t \mid \mathbf{z}(t); \beta)$ to relate the hazard function (risk of failure) to the covariate history at time t and the parameters β. Alternatively the cumulative hazard function can be used.

Note: Failure time models may always be represented as hazard models, but hazard models can only usually make predictions for failure time distributions

(a) when the covariates are *fixed*; or

(b) when the time variation of $\mathbf{z}(t)$ is totally *deterministic*.

When there are *stochastic covariates*, the failure time distribution cannot be predicted without predicting the covariate process.

Statistical methods based on using parametric models, with information from a random sample, are concerned with such questions as

(i) 'What is the best choice of the values of the parameters β, or some function of β, on the basis of the data observed?' (*estimation*); or

(ii) 'Are the data observed consistent with the parameters β having the value β_0, say?' (*hypothesis testing*).

There have been a variety of ways of answering the first question over the years, though it is generally accepted that there are essentially three main approaches: *classical* (due to Neyman and Pearson), *likelihood* (due to Fisher), and *Bayesian* (developed by the followers of Bayes). All three methods have their advocates and they will be briefly described in Chapter 2, Section 2.6. The *classical* and *likelihood* approaches answer the second question and will be described in Chapter 2, Section 2.7.

1.7.2   NON-PARAMETRIC MODELS

As an alternative to choosing a parametric failure time distribution it is possible to make no specification of the reliability or probability density function of this distribution. This has advantages and disadvantages. While not being restricted to use specific statistical methods, which may not be valid for the model being considered, the price to be paid for this is that it may be difficult to make very powerful inferences from the data. This is because a *non-parametric* model can be effectively thought of as a parametric model with an **infinity** of parameters. Hence with an infinity of parameters it is not surprising that statistical methods may be unable to produce powerful results. However, this approach to models and the resulting statistical methods which follow from it are often successful and hence have been very popular. Statistical methods of estimation for non-parametric models will be considered first in Chapter 2, Section 2.5, before methods of parametric estimation are considered in Section 2.6.

1.7.3   SEMI-PARAMETRIC MODELS

As a compromise to choosing a parametric or a non-parametric failure time distribution it is possible to make only a partial specification of the reliability, probability density, or hazard function of this distribution. This has the advantage of introducing parameters which model the covariates of interest while leaving the often uninteresting remainder of the model to have a non-parametric form. Such models are referred to as *semi-parametric* models. Probably the most famous example of a semi-parametric model is the hazard model introduced by Cox (1972), known as the *proportional hazards* model, which is described in Chapter 3, Section 3.5. The idea behind using such a model is to be able to use the standard powerful approach of parametric methods to make inferences about the covariate parameters while taking advantage of the generality of non-parametric methods to cope with any problems produced by the remainder of the model. How successfully this can be done is still an open question for many models, but this may become clearer in the future.

1.8 Discussion

The problems encountered in using methods of statistical data analysis in reliability studies need to be illustrated with reference to reliability case studies. For this reason three examples of case studies were given in Section 1.2, which will be used later in Chapter 10. These are taken from the authors' experience of applying statistical methods to problems concerning the reliability of equipment from a wide range of technologies.

It is important to emphasize, as was stated in Section 1.3, that statistical analysis is *not* performed without an objective in mind. Any reliability study will have been conducted with an objective in mind and this objective must dictate any statistical analysis that is to be performed. Sometimes none, or very little, may be needed. At other times the objective of the study can only be achieved by using the latest and most powerful statistical techniques available. The approach should always be 'problem-led' rather than 'technique-led'.

The three case studies described in Section 1.2 cannot cover all the techniques which are discussed, so these will be illustrated using a number of examples of different kinds of reliability data described in Section 1.4. These examples covered censoring, truncation, and repair.

Data of whatever kind have to be collected if they are to be analysed by statistical methods. So in Section 1.5 the problems associated with the use of historic and the creation of new databases were considered.

The formal definition of reliability is in terms of probability and in Section 1.6 the basic definitions and ideas were reviewed. The important concept of a random variable and its associated probability distribution was introduced for the simpler discrete case before the ideas of a failure time distribution were covered using the continuous case. The relationships between the reliability function, the probability density function, and the hazard function were detailed. The hazard function is especially important in reliability as many parametric models are defined in terms of this function. The negative exponential distribution was discussed as an example of a failure time distribution as its hazard function is constant.

The chapter was concluded with an account of statistical methods in Section 1.7, which covered the use of parametric, semi-parametric, and non-parametric models.

2
Lifetime distributions

2.1 Introduction

This chapter introduces the basic ideas of parametric lifetime distributions, by firstly defining expectation, the median, and the mode, and then giving details of the most important examples encountered in reliability. These are the gamma, Weibull, extreme value, and normal distributions. In all cases the probability density functions are illustrated and for the gamma and Weibull cases the hazard functions are illustrated. After a treatment of the various kinds of censoring mechanisms that can produce examples of censored data, the problem of estimation is considered from the non-parametric standpoint. The well-known Kaplan–Meier and Nelson–Altschuler estimators for right censored data are described. When using parametric lifetime distributions values of the parameters have to be chosen. The 'best' values have to be chosen (estimated) on the basis of the information from the data available. The various methods of estimating parameters in parametric models from the classical, likelihood, and Bayesian viewpoints are explained. The chapter is concluded with an account of hypothesis testing which is applied to the question of the goodness-of-fit of a hypothesized lifetime distribution for censored and truncated samples.

2.2 Properties of distributions

As was described in Chapter 1, Section 1.6, the distributions of continuous random variables are defined using the *reliability function* and the *probability density function*. If these functions are known for every value t taken by the random variable T then the probabilities of all events specified by T can be obtained. However, it may be desirable to summarize the probabilistic properties of the distribution. One method is to use *expectation* which is a measure of the location of the probability of the distribution. It is analogous to 'centre of gravity' of mass in mechanics.

Definition of expectation
The expectation of a function $g(T)$ of a continuous random variable T is

$$E(g(T)) = \int_0^\infty g(u) f_T(u) du.$$

If $g(T) = T$ then $E(T)$ is known variously as the 'expected value', 'mean (value)', or 'average (value)' of T, and is usually denoted by μ.

Expectation can also be used to give a 'measure of spread' of the probability of a distribution around μ. This can be done by using $g(T) = (T-\mu)^2$, where $E((T - \mu)^2)$ is known as the *variance* and is usually denoted by σ^2. This quantity is *non-negative* and its positive square root is known as the *standard deviation* and is denoted by σ.

Note: In the case of discrete distributions described in Chapter 1, Subsection 1.6.2, the definition of expectation involves summing the probability function rather than using an integral of the probability density function. Hence $\mu = np$ and $\sigma^2 = np(1-p)$ for the binomial distribution in Example 1.6.1 and $\mu = \sigma^2 = \mu$ for the Poisson distribution in Example 1.6.2, which explains the choice of symbol for the parameter in the Poisson case.

There are two other popular measures used to summarize the probabilistic properties of a distribution. Firstly the *median* or 'middle value' which is the value of t for which the reliability function is a half. So the median, t_m, satisfies

$$R_T(t_m) = 0.5.$$

Secondly the *mode* is the value of t at which the probability density function has a local maximum value, i.e. the derivative of the function changes from positive to negative values. So for a distribution with one mode this is the value around which there is the highest density of probability.

2.3   Examples: exponential, Weibull, extreme value, normal

Examples of four parametric families of distributions for continuous random variables will now be given.

2.3.1   EXPONENTIAL AND GAMMA DISTRIBUTIONS

A continuous random variable T has an exponential distribution if its probability density function is given by

$$f_T(t) = \begin{cases} \lambda \exp(-\lambda t), & \text{for } t \geq 0, \\ 0, & \text{for } t < 0, \end{cases}$$

for $\lambda > 0$. As was shown in Example 1.6.3 in Chapter 1 the parameter λ is the value of the hazard function which is **constant**, for all t. However, λ is

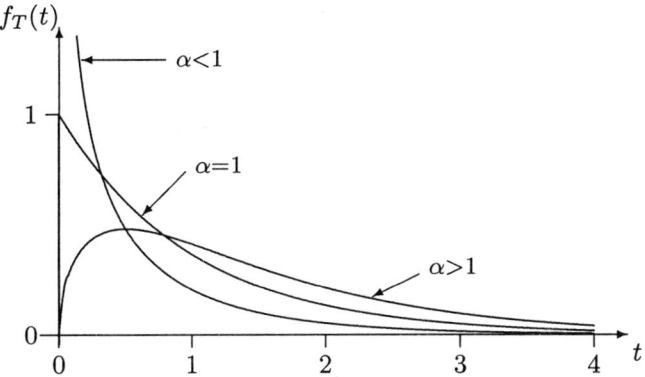

Fig. 2.1.   Gamma probability density functions for $\lambda = 1$ and various α

also the reciprocal of both the expected value (mean), μ, and the standard deviation σ.

The connection between this distribution and the Poisson distribution, as stated in Example 1.6.3 in Chapter 1, can now be further explained using expectation. The exponential distribution is the distribution of the times between events which are from a Poisson process and the mean of these times is the reciprocal of λ. However, λt is the mean of the Poisson distribution for the number of events in a fixed interval of time of length t and hence λ is the 'rate at which events occur'. So the mean time between events is the reciprocal of the rate at which events occur for a Poisson process. The exponential distribution is a special case of the **gamma** distribution.

A continuous random variable T has a gamma distribution if its probability density function is given by

$$f_T(t) = \begin{cases} \lambda^\alpha t^{\alpha-1} \exp\left(-\lambda t\right)/\Gamma(\alpha), & \text{for } t \geq 0, \\ 0, & \text{for } t < 0, \end{cases}$$

for $\alpha > 0$ and $\lambda > 0$, where α is a *shape* parameter, λ is a *scale* parameter, and $\Gamma(\alpha)$ is the gamma function. The gamma function is defined by the integral

$$\Gamma(\alpha) = \int_0^\infty x^{\alpha-1} \exp\left(-x\right) dx, \text{ for } \alpha > 0.$$

(When $\alpha = 1$ the distribution is the exponential distribution.) The expected value is α/λ, the variance is α/λ^2, and the mode is $(\alpha - 1)/\lambda$, if $\alpha > 1$. The probability density function is illustrated for $\lambda = 1$ and various α in Fig. 2.1.

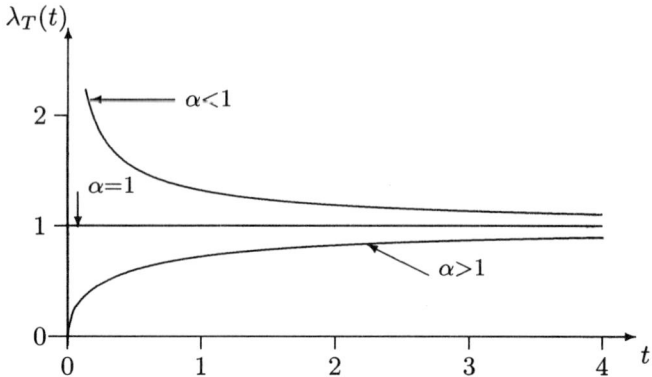

Fig. 2.2. Gamma hazard functions for $\lambda = 1$ and various α

The reliability function is given by

$$R_T(t) = \begin{cases} \Gamma(\alpha, \lambda t)/\Gamma(\alpha), & \text{for } t \geq 0, \\ 1, & \text{for } t < 0, \end{cases}$$

where $\Gamma(\alpha, y)$ is the incomplete gamma function which is defined by the integral

$$\Gamma(\alpha, y) = \int_y^\infty x^{\alpha-1} \exp(-x)dx, \text{ for } \alpha > 0 \text{ and } y > 0.$$

Hence the hazard function is given by

$$\lambda_T(t) = \begin{cases} \lambda^\alpha t^{\alpha-1} \exp(-\lambda t)/\Gamma(\alpha, \lambda t), & \text{for } t \geq 0, \\ 0, & \text{for } t < 0. \end{cases}$$

The hazard function is illustrated for $\lambda = 1$ and various α in Fig. 2.2. It can be seen that the hazard function is monotonic and mirrors the behaviour of the probability density functions for small t. For large t all hazard functions approach the same limit of λ.

The gamma distribution is also connected to the exponential distribution because the sum of n independent random variables, which all have the exponential distribution with the same hazard function λ, will have a gamma distribution with scale parameter λ and shape parameter $\alpha = n$. Hence, when $\alpha = n$, the distribution arises in connection with the Poisson process as the distribution of the time between n events. It is also known as the chi-square distribution with ν degrees of freedom (χ_ν^2), when $\lambda = \frac{1}{2}$ and $\alpha = \frac{\nu}{2}$, where ν is a positive integer.

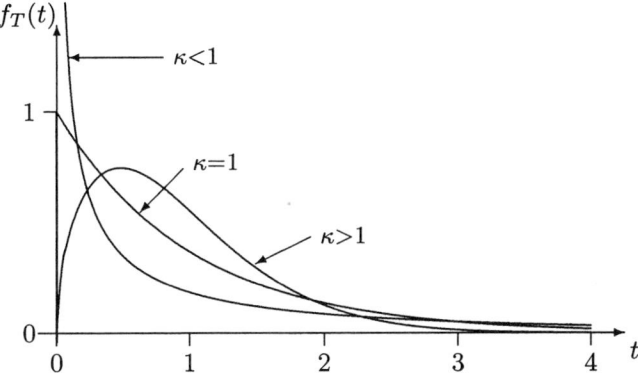

Fig. 2.3.   Weibull probability density functions for $\theta = 1$ and various κ

2.3.2   WEIBULL DISTRIBUTION

A continuous random variable T has a Weibull distribution if its probability density function is given by

$$f_T(t) = \begin{cases} \kappa t^{\kappa-1} \exp\left(-(t/\theta)^\kappa\right)/\theta^\kappa, & \text{for } t \geq 0, \\ 0, & \text{for } t < 0, \end{cases}$$

for $\kappa > 0$ and $\theta > 0$, where κ is a *shape* and θ is a *scale* parameter. The probability density function is illustrated for $\theta = 1$ and various κ in Fig. 2.3. Using the gamma function defined in Subsection 2.3.1, the expected value is $\theta\Gamma(1 + 1/\kappa)$ and the variance is $\theta^2\{\Gamma(1 + 2/\kappa) - \Gamma(1 + 1/\kappa)^2\}$. The mode is $\theta(1 - 1/\kappa)^{1/\kappa}$ for $\kappa > 1$ and the median is $\theta(\log 2)^{1/\kappa}$. Again the exponential distribution is a special case of this distribution, when $\kappa = 1$ and the hazard function is constant $(= 1/\theta)$.

In some applications a third (usually positive) parameter η is introduced. This parameter replaces zero, the lower limit of the interval on which the probability density function $f_T(t)$ is positive, and is referred to as the *threshold* (or guarantee) parameter. This parameter is used when the failure time cannot occur before a certain time, the value of this third parameter. The Weibull distribution was popularized by Weibull (1951) and is widely used in reliability data analysis as it has been found useful to model the failure times of many kinds of components. It has a simple reliability function and can model a variety of hazard functions. The reliability function is given by

$$R_T(t) = \begin{cases} \exp\left(-(t/\theta)^\kappa\right), & \text{for } t \geq 0, \\ 1, & \text{for } t < 0, \end{cases}$$

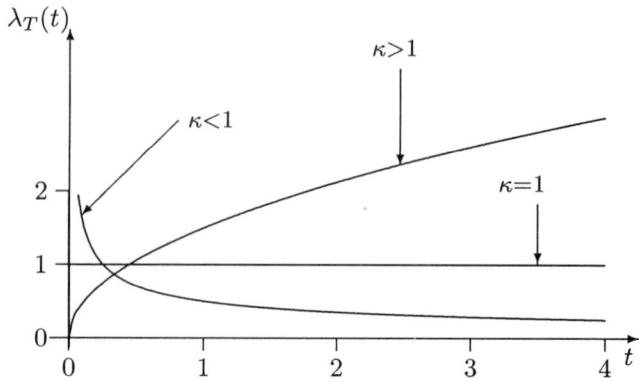

Fig. 2.4. Weibull hazard functions for $\theta = 1$ and various κ

and the hazard function is given by

$$\lambda_T(t) = \begin{cases} \kappa t^{\kappa-1}/\theta^\kappa, & \text{for } t \geq 0, \\ 0, & \text{for } t < 0. \end{cases}$$

So the shape parameter κ determines whether the hazard function is an increasing ($\kappa > 1$), a constant ($\kappa = 1$), or a decreasing ($\kappa < 1$) function of time t. The hazard function is illustrated for $\theta = 1$ and various κ in Fig. 2.4.

However, the hazard function is always *monotonic*, so that it either always decreases or increases. So it is not possible to use the Weibull distribution to model situations where the hazard function either increases and then decreases or decreases and then increases.

2.3.3   EXTREME VALUE DISTRIBUTION

The extreme value distribution arises as one of the possible limiting distributions obtained when the minima of a number of variables are considered. It would therefore represent the time to the first failure if a component could fail from a number of different possible causes. It can also be obtained, by reversing the direction of the time axis, as the maximum of a number of variables and hence is the most 'extreme' value obtained from the variables. Hence the reason for its name, though it is also associated with Gumbel (1960) and Gompertz, and is referred to by these names in some texts. It is defined for negative and positive values and so is not generally used as a lifetime distribution as there is a positive probability of negative values.

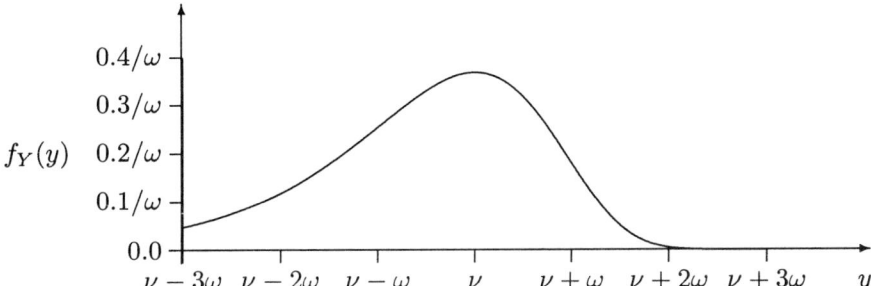

Fig. 2.5.   Extreme value probability density function

A continuous random variable Y has an extreme value distribution if its probability density function is given by

$$f_Y(y) = \frac{1}{\omega} \exp\left(\frac{y - \nu}{\omega} - \exp\left(\frac{y - \nu}{\omega}\right)\right), \text{ for } -\infty < y < \infty,$$

for ν, the mode, and $\omega > 0$. The expectation is $\nu - \omega\gamma$, where $\gamma = 0.5772$ is Euler's constant, and the variance is $\omega^2\pi^2/6$. The probability density function is illustrated in Fig. 2.5. When $\nu = 0$ and $\omega = 1$ the distribution is known as the standard extreme value distribution.

The reliability function is given by

$$R_Y(y) = \exp\left(-\exp\left(\frac{y - \nu}{\omega}\right)\right), \text{ for } -\infty < y < \infty.$$

Hence the hazard function is always an increasing function of y.

This distribution is closely connected with the Weibull distribution because if $\nu = \log\theta$ and $\omega = 1/\kappa$ then the random variable $\exp(Y)$ has the Weibull distribution with parameters θ and κ given in Subsection 2.3.2. Or, alternatively, a random variable $\log T$ has an extreme value distribution if T has a Weibull distribution.

2.3.4   NORMAL (GAUSSIAN) DISTRIBUTION

The normal distribution is an important distribution used in the analysis of data. It is defined for negative and positive values, so like the extreme value distribution it is not generally used as a lifetime distribution as there

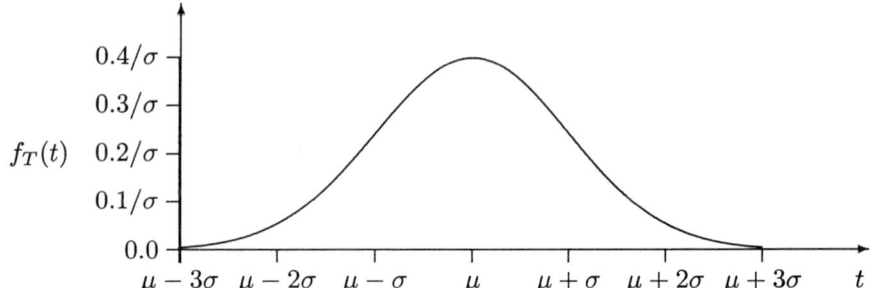

Fig. 2.6. Normal probability density function for μ and σ

is a positive probability of negative values. It is possible to choose the parameters suitably so that the probability of negative values is negligible. A theoretical justification for using a normal distribution comes from the central limit theorem when the variable is considered to be the sum of a large number of other variables. This distribution plays a central part in the theory of statistical methods, and is associated with K.F. Gauss.

A continuous random variable Y has a normal (or Gaussian) distribution if its probability density function is given by

$$f_Y(y) = \frac{1}{\sqrt{2\pi}\sigma} \exp\left(-\frac{(y-\mu)^2}{2\sigma^2}\right), \text{ for } -\infty < y < \infty,$$

for μ, which is the expected value, and $\sigma > 0$, which is the standard deviation. It is symmetric about μ, which is the value of the median and mode as well as the expected value (mean). The probability density function is illustrated in Fig. 2.6. There is no easy form for the reliability function which is evaluated numerically and found in standard statistical tables. In reliability applications the normal distribution is used like the extreme value distribution given in Subsection 2.3.3 for the distribution of a random variable $\log T$. Then the random variable T has a *lognormal* distribution. The expected value is $\exp(\mu + \frac{1}{2}\sigma^2)$ and the variance is $\exp(2\mu + \sigma^2)\{\exp(\sigma^2) - 1\}$, where μ and σ are the mean and standard deviation of the normal distribution. The mode is $\exp(\mu - \sigma^2)$ and the median is $\exp(\mu)$.

2.4   Censored and truncated samples

As has been described in Chapter 1, Section 1.4, it is not always possible to collect data for lifetime distributions in as complete a form as required. The example given in Subsection 1.4.1 is an example of a complete sample where a failure time was observed for every system. This situation is unusual and only likely to occur in the case of laboratory testing. Many studies lead to the collection of incomplete data which are either censored or truncated. One of the most frequent ways that this occurs is because of a limit of time (or resources) for the study, which produces the case of right censored data. This case will be considered first.

2.4.1   CENSORED DATA

Observation of a system may cease before the event of interest has occurred at time T. This is called *right censoring*, as observation is not complete on the *right* (positive) end of the conventional time axis. The time at which observation ceases is denoted by C. It is conventional to record the ith event time in such problems as a pair, (t_i, d_i), where

$$t_i = \mathrm{Min}(T, C),$$

and

$$d_i = \left\{ \begin{array}{ll} 1, & \text{if failure is observed (at } t_i = T), \\ 0, & \text{otherwise.} \end{array} \right.$$

The variable d_i is known as a *censoring indicator*.

Example 2.4.1
An example of right censored failure data was presented in Chapter 1, Subsection 1.4.2. As was seen in that example the number of censored observations may be a high proportion of the total event times, as in that case the number was 87% of the total. This illustrates the fact that in practice censoring cannot be ignored.

□

A number of definitions of right censoring are now presented.
(i) *Fixed censoring*
C is *fixed*, so it is determined at the start of the study.
(ii) *Random censoring*
C is a *random variable* with some distribution with a probability density function, $f_C(c)$, say.

The important special case in which valid inferences can be made about the parameters β of a failure time distribution using right censored data is now given.

(iii) *Independent censoring*
C and T are independent random variables conditional upon the covariates
$\mathbf{z}$ so that the product rule holds for the conditional probability density
functions with

$$f_{T,C|\mathbf{Z}}(t, c \mid \mathbf{z}; \beta) = f_{T|\mathbf{Z}}(t \mid \mathbf{z}; \beta) f_{C|\mathbf{Z}}(c \mid \mathbf{z}).$$

Right censored data are the most frequently met case of observations
which have been censored. Left censored data are observed in the case
when the observation of failures is not started until after an initial period.
So only failures after this period are observed, otherwise it is known that
the failure occurred in the initial period. Again a censoring indicator can
be used to record the event times. This case is not as frequent as right
censoring, and is probably due to poor planning when it occurs. If it is
known only that the failure occurred between two times, rather than the
exact time, then this is known as interval censoring.

2.4.2   TRUNCATED DATA

On the other hand *truncation* is the case when the time of failure of a
system T is not observed if it fell in a certain interval of time. The most
frequent case is when the data are not collected from the beginning of a
study but at some later time after the start of using the systems.

Example 2.4.2
An example of truncated failure data for compressors in heat pumps was
presented in Chapter 1, Subsection 1.4.3. In that example observations
were only collected during the period when the building was on service
contract and when this period started only after the compressors had been
in service for a time (2.59 years in the example) so there is no information
in this initial period. This is an example of *left* truncation. The service
contract lasted until a time of 9.33 years when observation of the 164 compressors ceased. During this period 10 failure times were observed so the
remaining 154 compressors had right censored observations at 9.33 years.
This illustrates the fact that in practice truncation and censoring may both
be features of the failure data.

□

2.5   Non-parametric estimation of distributions of failure times

Non-parametric estimation methods have been mainly developed to deal
with the case of right censored data for lifetime distributions, as this is the
case most widely encountered in all branches of applied statistics.

Consider right censored failure time data for n components (or systems). Such data can be represented by the pairs, (t_i, d_i), for the ith component, $i = 1, 2, ..., n$. For the most part the case in which the T_i are independent and identically distributed is considered and where it is necessary to estimate some characterization of their distribution.

Sometimes it is necessary to refer to the ith ordered observed *event time* (the time of failure or censoring of the ith event) when the notation $t_{(i)}$ (rather than t_i) will be used. Then

$$t_{(1)} \le t_{(2)} \le t_{(3)} \le ..., \text{ etc.}$$

On the other hand it is sometimes necessary to refer to the ith observed failure time (rather than the time of the ith event, which could be a failure or censoring). Then the notation $t_{[i]}$ (rather than t_i) will be used. Thus

$$t_{[1]} < t_{[2]} < t_{[3]} < ..., \text{ etc.}$$

Corresponding to each observed failure time is a *risk set*, the set comprising those components who were under observation at that time, i.e. the set of components who *could* have been the observed failure. The risk set corresponding to $t_{[i]}$ will be denoted by R_i. The number of components in the risk set R_i will be denoted by r_i. These definitions are illustrated in Example 2.5.3 in Table 2.1.

Non-parametric estimators have been proposed for the reliability (survivor) function and the cumulative hazard function of the failure time distribution. These will now be presented and applied to examples of failure data.

2.5.1 THE KAPLAN–MEIER (KM) ESTIMATOR OF THE RELIABILITY FUNCTION

The KM estimator $\hat{R}_T(t)$ of the reliability function, $R_T(t)$, which was justified by Kaplan and Meier (1958), is a step function given by

$$\hat{R}_T(t) = \prod_{i:t_{[i]} \le t} \left\{ 1 - \frac{1}{r_i} \right\}.$$

In the case of ties of multiplicity m_i at the failure time $t_{[i]}$ the estimator is modified and is given by

$$\hat{R}_T(t) = \prod_{i:t_{[i]} \le t} \left\{ 1 - \frac{m_i}{r_i} \right\}.$$

If censoring times tie with failure times, censoring is assumed to occur *just following the failure.*

The KM estimator gives a 'point' estimator, or single value for the reliability function at any time t. If it is desired to obtain a measure of the variation of this estimator over different samples then an estimate of the variance of the KM estimator is needed, and this is now given.

Greenwood's formula provides an estimate of the variance of $\log(\hat{R}_T(t))$ given by

$$\widehat{\mathrm{Var}}(\log(\hat{R}_T(t))) = \sum_{i:t_{[i]} \leq t} \left\{ \frac{m_i}{r_i(r_i - m_i)} \right\}.$$

Though this gives a measure of the variation of the log of the KM estimator, it is possible to use it to estimate the variance of $\hat{R}_T(t)$ by $(\hat{R}_T(t))^2 \widehat{\mathrm{Var}}(\log(\hat{R}_T(t)))$.

2.5.2 THE NELSON–ALTSCHULER (NA) ESTIMATOR OF THE CUMULATIVE HAZARD FUNCTION

The NA estimator $\hat{\Lambda}_T(t)$ of the cumulative hazard function, $\Lambda_T(t)$, which was proposed by Nelson (1969), is also a step function and is given by

$$\hat{\Lambda}_T(t) = \sum_{i:t_{[i]} \leq t} \left\{ \frac{1}{r_i} \right\}.$$

There is some difficulty with ties in the theory of this estimator, it having been derived in the case of *continuous time*. One fairly common variant is

$$\hat{\Lambda}_T(t) = \sum_{i:t_{[i]} \leq t} \left\{ \frac{m_i}{r_i} \right\}.$$

Ties between failure times and censoring times are dealt with as in Subsection 2.5.1.

The asymptotic variance estimate is

$$\widehat{\mathrm{Var}}(\hat{\Lambda}_T(t)) = \sum_{i:t_{[i]} \leq t} \left\{ \frac{m_i}{r_i^2} \right\}.$$

Using the basic relation between the reliability function and the cumulative hazard function given in Chapter 1, Subsection 1.6.3, it is possible to use the NA estimator to estimate the reliability function by

$$\hat{R}_T(t) = \exp\left(-\hat{\Lambda}_T(t)\right).$$

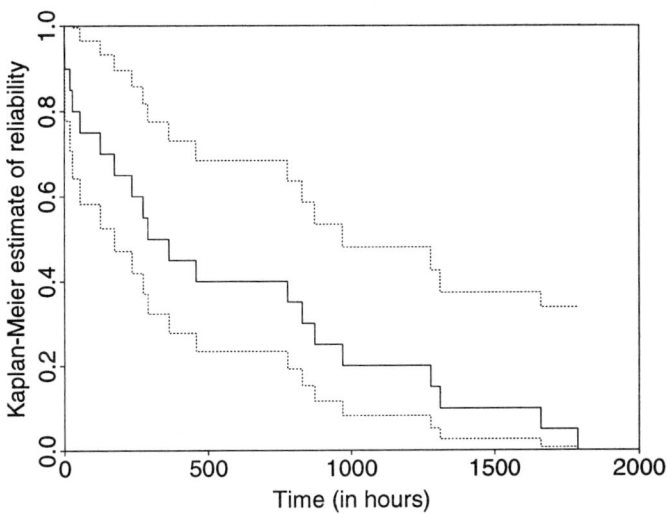

Fig. 2.7. Plot of the Kaplan–Meier estimator of the reliability for the pressure vessels' laboratory test data with 95% confidence limits

2.5.3   EXAMPLES

A number of examples are now presented to illustrate the features of these non-parametric estimators of the reliability function.

Example 2.5.1
An example of a single sample from a laboratory test of 20 failure times for pressure vessels was presented in Chapter 1, Subsection 1.4.1. In that example all the observations were distinct failure times so that the sample was complete and there was no truncation or censoring. In this case the KM estimator is a step function with equal steps of 0.05 (=1/20) at the 20 failure times. A plot of the KM estimator is given in Fig. 2.7 with 95% confidence limits obtained using Greenwood's formula.

□

Example 2.5.2
An example of a sample with right censored failure times for 125 telecommunication systems was presented in Chapter 1, Subsection 1.4.1. In that example the observations consisted of 16 failure times and 109 right censored observations. In this case the KM estimator is a step function with equal steps of 0.008 (=1/125) for the first 11 failure times. After that

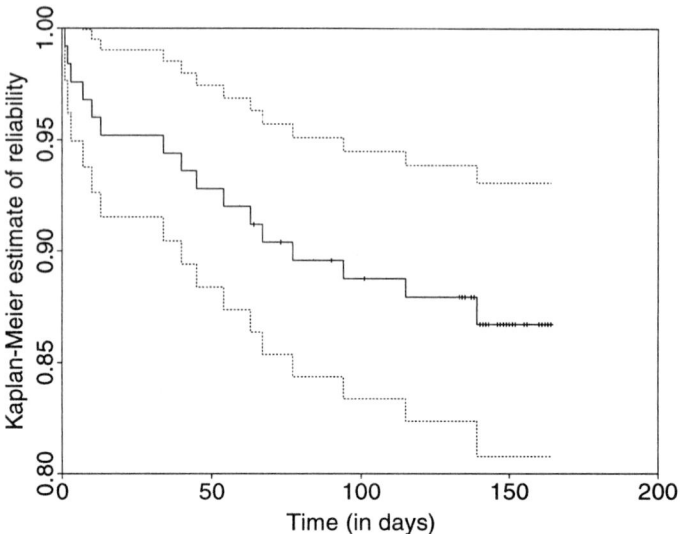

Fig. 2.8. Plot of the Kaplan–Meier estimator of the reliability for the telecommunication systems data with 95% confidence limits

time (63 days) right censored observations occur and the step size changes until the largest failure time (139 days) is reached. After that time the KM estimator remains constant until the largest censored time (164 days), after which the estimator is not defined. So in this example, unlike Example 2.5.1, the KM estimator is never zero. A plot of the KM estimator is given in Fig. 2.8 with 95% confidence limits obtained using Greenwood's formula. (The vertical lines on the plot of the reliability represent the censored times.)

$\square$

Example 2.5.3
This example is adapted from Newton (1991), from the data which were described in Chapter 1, Subsection 1.4.5. Five copies of a system were put into operation at 500 hour intervals. (For the purpose of this example the system will be considered to be equivalent to a single component, and will be referred to as a component.) When failures occurred the component was instantaneously replaced by a new component. Failures were logged at cumulative system hours and are given in Table 1.7 except that the failure time of system D has been changed from 782 to 783, for illustrative purposes, as this will produce two tied failure times and a multiplicity of

Table 2.1. Ordered event times (in hours)

Event number (i)	Censor indicator $d_{(i)}$	Event time $t_{(i)}$	Fail number $[i]$	Multiplicity $m_{[i]}$	Risk set $r_{[i]}$	$\hat{\Lambda}(t_{[i]})$	$\hat{R}(t_{[i]})$ NA	$\hat{R}(t_{[i]})$ KM
1	1	69	1	1	21	0.048	0.953	0.952
2	1	176	2	1	20	0.098	0.907	0.905
3	0	196+	-	-	-	-	-	-
4	1	208	3	1	18	0.153	0.858	0.854
5	1	215	4	1	17	0.212	0.809	0.804
6	1	233	5	1	16	0.274	0.760	0.754
7	1	289	6	1	15	0.341	0.711	0.704
8	1	300	7	1	14	0.413	0.662	0.653
9	1	384	8	1	13	0.490	0.613	0.603
10	1	390	9	1	12	0.573	0.564	0.553
11	0	393+	-	-	-	-	-	-
12	1	401	10	1	10	0.673	0.510	0.498
13	1	452	11	1	9	0.784	0.457	0.442
14	1	567	12	1	8	0.909	0.403	0.387
15	0	617+	-	-	-	-	-	-
16	0	718+	-	-	-	-	-	-
17–18	1	783	13	2	5	1.309	0.270	0.232
19	1	806	14	1	3	1.642	0.194	0.155
20	0	1000+	-	-	-	-	-	-
21	1	1022	15	1	1	2.642	0.071	0.000

Note: Times with + are right censored times.

2. The details of the calculation of both the KM and NA estimators will now be illustrated.

For the analysis of the distribution of the component failure times the observed lifetimes (times between failures) are ranked as shown in Table 2.1. The events with + in Table 2.1 are times from the last failure of a component to the current time. Such times are right censorings (times to non-failure) and are typical of failure data from field service. As most reliability data come from such sources, it is important **not** to analyse just the failure times and ignore the censoring times by treating the data as if they came from a complete sample as in a laboratory test. Ignoring censorings will result in pessimistic estimates for reliability. Data are often encountered, as in Example 2.5.2, where there are as much as four or five times as many censorings as failures. The consequences of ignoring the censorings would result in wildly inaccurate conclusions about the lifetime distribution.

An estimate can be given for the reliability function at each observed component failure time. The estimate of the reliability function in the last column of Table 2.1 is that obtained by using the KM estimator described in Subsection 2.5.1.

Alternatively the NA estimator can be used. The estimate of the hazard function at each failure time is obtained as the number of failures (the multiplicity $m_{[i]}$) at each event time divided by the number of survivors immediately prior to that time (the number in the risk set, $r_{[i]}$). The cumulative sum of these values is the NA estimate, $\hat{\Lambda}(t)$, of the cumulative hazard function $\Lambda(t)$, described in Subsection 2.5.2. Then $\exp(-\hat{\Lambda}(t))$ provides an estimate of $R(t)$, the reliability function, and is given in the penultimate column of Table 2.1. It can be seen that the NA estimate of the reliability function is consistently slightly higher than the KM estimate.

The improbability of tied lifetimes should lead to consideration of the possibility of the times not being independent, as would be the case in, for example, common cause failures, see Chapter 8, Section 8.4. Both these estimates have been derived assuming that the lifetimes are independent and identically distributed (IID).

$\square$

2.6   Methods for the estimation of parameters

To achieve the objectives of the reliability study it may be decided to use a parametric model with parameters β, say. This may be to obtain the reliability of the system or the expected value of the failure time, and in order to do this it will be necessary to specify the values of the parameters of the model. This will be done by using the 'best' value obtained from the reliability data, which are the observations collected (in a possibly censored or truncated form) from a random sample. *Estimation* is the process of choosing a function of these observations (known as a *statistic*) which will give a value (or values) close to the 'true' (unknown) value of the parameters β. Such a statistic, known as an *estimator*, will be a random variable denoted by $\hat{\beta}$. A *point* estimator gives one value for each of the parameters, whilst an *interval* estimator gives intervals for each of the parameters. For interval estimators a probability (or *confidence level*, usually expressed as a percentage by multiplying the probability by 100) is specified, which is the probability that the interval contains the 'true' values of the parameter. One method for obtaining the interval is to use the *standard error* of an estimator, which is the standard deviation of the estimator with any parameters replaced by their estimates. Often the estimator can be assumed to have a normal distribution (at least for a large sample size, though this depends on the proportion of censored observations). Then an approximate 95% interval estimator for β using the point estimator $\hat{\beta}$ with a standard error $se(\hat{\beta})$ would be given by the interval $(\hat{\beta} - 1.96se(\hat{\beta}), \hat{\beta} + 1.96se(\hat{\beta}))$. Whether an estimator is 'good' or not will depend on its properties. Four properties which are considered desirable for point estimators will now be outlined.

2.6.1   PROPERTIES OF ESTIMATORS

The following four properties have been traditionally advocated as desirable for point estimators. The estimator for the parameter β will be denoted by $\hat{\beta}$ and will be considered as a single random variable for ease of explanation. The ideas are easily generalized to more variables.

Bias

An estimator $\hat{\beta}$ is *unbiased* if its expected value, $E(\hat{\beta})$, is equal to the parameter β. Otherwise the estimator has a *bias* equal to the difference between its expected value and the value of the parameter.

Consistency

An estimator $\hat{\beta}$ is *consistent* if the probability that it differs by a constant value from its expected value decreases to zero as n, the sample size, increases. This will occur if the bias and variance of $\hat{\beta}$ both decrease to zero as n increases. So as more and more information is obtained by sampling, the estimator approaches closer to the true value of the parameter.

Efficiency

The relative *efficiency* of two unbiased estimators is the ratio of the inverses of their variances. An *absolute* measure of efficiency is obtained relative to an unbiased estimator of *minimum variance*, known as an *efficient* estimator. A lower bound on this minimum variance may be obtained from the *information* of the sample. This is known as the *Cramer–Rao* or *Frechet–Darmois* lower bound, though it is not always attained.

Sufficiency

An estimator $\hat{\beta}$ is a *sufficient* statistic for the parameter β if it contains all the information contained in the sample. This is equivalent to the statement that the conditional distribution of the sample given the sufficient statistic does not depend on the parameter β.

Mean square error

The mean square error of an estimator $\hat{\beta}$ is the expectation of the square of the difference between $\hat{\beta}$ and the parameter β. It measures how 'far away' the estimator is from the parameter β 'on average'.

2.6.2 CLASSICAL METHODS FOR THE DETERMINATION OF ESTIMATORS

One of the oldest methods of determining estimators is the *method of moments* (associated with K. Pearson). It is necessary to give two definitions for *moments* to understand this method.

Population moments

The ith *population moment* of the random variable T, say, is defined to be $E(T^i)$. So the expectation μ is the first population moment and the variance σ^2 is defined in terms of the second population moment.

Sample moments

The ith *sample moment* for a sample $t_1, t_2, ..., t_n$ of size n is defined to be

$$\tilde{\mu}_i = \sum_{j=1}^{n} t_j^i / n.$$

(So the first sample moment $\tilde{\mu}_1 = \bar{t}$, the sample mean.)

With these two definitions it is now possible to outline the method.

Method of moments

If there are k parameters to be estimated then the first k population moments are equated to the corresponding sample moments. The solutions of the k equations give the estimators of the parameters, which will be functions of the sample moments. These estimators will generally be consistent and though their efficiency is often less than one they may be used as first approximations from which more efficient estimates may be obtained.

Example 2.6.1

For the exponential parametric distribution, given in Subsection 2.3.1, the expected value is $1/\lambda$. Equating this expectation to the first sample moment, $\bar{t}$, so that $1/\hat{\lambda} = \bar{t}$, the estimator $\hat{\lambda}$ for λ is given by

$$\hat{\lambda} = \frac{1}{\bar{t}},$$

which is the inverse of the sample mean of the observed failure times.

Using the example of failure times for the 20 pressure vessels presented in Subsection 1.4.1 the estimate of λ is $1/575.3 = 0.00174$.

□

It is often convenient for simplicity to consider only those estimators which are linear functions of the sample observations.

Best linear unbiased estimators
Among those unbiased estimators which are linear functions of the sample observations the *best linear unbiased estimator* is the estimator with minimum variance.

Example 2.6.2
For the binomial parametric distribution, given in Examples 1.6.1 and 1.7.2, the parameter p is the probability that a system fails. Let X_i be

$$X_i = \begin{cases} 1, & \text{if failure is observed,} \\ 0, & \text{otherwise.} \end{cases}$$

Then for a sample of size n a linear estimator is given by

$$\hat{p} = \sum_{j=1}^{n} c_j X_j,$$

where the c_j are suitably chosen constants. If this estimator is unbiased then $\sum_{j=1}^{n} c_j = 1$. The variance is given by $pq \sum_{j=1}^{n} c_j^2$ and it can be shown that this variance is minimized if $c_j = 1/n$, for $j = 1, \ldots, n$. Hence the best linear unbiased estimator is given by

$$\hat{p} = \bar{X}.$$

This is a point estimator.

Again using the example of failure times for the 20 pressure vessels presented in Subsection 1.4.1, consider finding the estimate of p, the probability that the failure time of a pressure vessel is less than 100 hours. Hence the point estimate is $5/20 = 0.25$.

□

Least squares estimators
The *least squares* estimator of a parameter β is the estimator which minimizes the sum of the squares of the differences between the observations of a sample and their expectations, which are a function of the parameter β. There are two reasons for using squares of the differences. Firstly, the square is a function which is non-negative and when minimized cannot be smaller than zero. Secondly, this will imply that large differences between an observation and its expectation will be penalized. Usually each observation is equally weighted, but there are often good reasons for using different weights, which will then give a *weighted least squares* estimator.

Example 2.6.3

For the binomial parametric distribution, considered in Example 2.6.2, the parameter p is the probability that a system fails. Then $E(X_i) = p$ and the least squares estimator of p will minimize

$$\sum_{i=1}^{n}(x_i - p)^2.$$

Hence, by differentiation by p, the least squares estimator is also found to be given by

$$\hat{p} = \bar{x}.$$

□

As will be shown in Subsection 2.6.3, when likelihood methods are considered, the least squares estimator is also the maximum likelihood estimator when a random sample is taken from a *normal population*.

2.6.3   LIKELIHOOD METHODS

The second method of determining estimators which is widely used is the *likelihood method*, which was propounded by R.A. Fisher. This method uses the *likelihood function*, which is the probability density function (probability function) of the joint distribution of the observations of the sample for a continuous (discrete) random variable, considering this function as a function of the parameter β. In the case of a random sample from a random variable T with probability density function $f_T(t; \beta)$, say, where the observations are IID, the likelihood function $L(\beta)$ is given by

$$L(\beta) = \prod_{i=1}^{n} f_T(t_i; \beta).$$

The principle of maximum likelihood

An estimator is obtained using the *principle of maximum likelihood* by choosing the value of β which maximizes the likelihood function $L(\beta)$. So for a fixed sample the *maximum likelihood estimator* maximizes the probability of observing that sample over all possible values of the parameter.

If the sample is from a normal population then using the probability density function given in Subsection 2.3.4 gives a likelihood function which is the exponential function of minus the sum of squares of the differences between the observations and their expectations. Hence the maximum likelihood estimator and the least squares estimator for any parameter used to model the expectation are equivalent in the case of a normal population.

Example 2.6.4
For the exponential parametric distribution, given in Subsection 2.3.1, the likelihood function for a sample of size n is

$$L(\lambda) = \lambda^n \exp\left(-\lambda \sum_{i=1}^{n} t_i\right).$$

When this likelihood is maximized as a function of λ this gives the estimator $\hat{\lambda} = 1/\bar{t}$, as was obtained in Example 2.6.1, by the method of moments. So using the example of failure times for the 20 pressure vessels presented in Subsection 1.4.1 the estimate of λ is again $1/575.3 = 0.00174$.

$\square$

The maximization of the likelihood function is often accomplished by using differential calculus to obtain appropriate equations which can then be solved. This maximization can often be more easily performed on the logarithm of the likelihood (*log likelihood*) function and this will be denoted by

$$l(\beta) = \log L(\beta).$$

Example 2.6.4 assumed that all the observations were failure times. However, as was illustrated in Section 2.4 many practical situations result in the collection of right censored observations. Hence the likelihood function given above for a complete random sample must be modified by replacing the probability density function by the reliability function for the right censored observations. Using the log likelihood function and the censoring indicator d_i introduced in Subsection 2.4.1 gives

$$l(\beta) = \sum_{i=1}^{n} d_i \log f_T(t_i; \beta) + \sum_{i=1}^{n} (1 - d_i) \log R_T(t_i; \beta).$$

Example 2.6.5
Consider again the example of Newton (1991) which was used in Example 2.5.3 to illustrate the estimation of the reliability function for component failure times by non-parametric methods. Suppose a parametric model is to be used to model the component failure time distribution in order to investigate the behaviour of the hazard function. If the Weibull distribution defined in Subsection 2.3.2 is chosen as a model then the log likelihood function is given by

$$l(\kappa, \theta) = d \log \kappa + (\kappa - 1) \sum_{i=1}^{n} d_i \log(t_i) - \sum_{i=1}^{n} (t_i/\theta)^\kappa - d\kappa \log \theta,$$

for $d = \sum_{i=1}^{n} d_i$ failure times and $n - d$ censored times. The partial differentials of this log likelihood function are given by

$$\frac{\partial l}{\partial \theta} = \frac{\kappa \sum_{i=1}^{n} t_i^{\kappa}}{\theta^{\kappa+1}} - \frac{d\kappa}{\theta}$$

$$\frac{\partial l}{\partial \kappa} = \frac{d}{\kappa} + \sum_{i=1}^{n} d_i \log(t_i) - \frac{\sum_{i=1}^{n} t_i^{\kappa} \log(t_i/\theta)}{\theta^{\kappa}} - d \log \theta.$$

The maximum likelihood estimators are the solutions of the two equations obtained by equating the two partial derivatives to zero. Then the parameter estimator $\hat{\kappa}$ is the solution of the equation in κ given by

$$\frac{\sum_{i=1}^{n} t_i^{\kappa} \log(t_i)}{\sum_{i=1}^{n} t_i^{\kappa}} - \frac{1}{\kappa} - \frac{\sum_{i=1}^{n} d_i \log(t_i)}{d} = 0.$$

Hence, by solving the above equation numerically, the shape parameter κ is estimated as $\hat{\kappa} = 1.662$ using the maximum likelihood estimation method for Example 2.5.3. The scale parameter, or characteristic life, estimator $\hat{\theta}$ is given by

$$\hat{\theta} = \left(\frac{\sum_{i=1}^{n} t_i^{\hat{\kappa}}}{d} \right)^{\frac{1}{\hat{\kappa}}},$$

and hence the characteristic life is estimated as $\hat{\theta} = 619.2$. These equations for the Weibull distribution were first given by Cohen (1965).

As the hazard function of a Weibull distribution is $\kappa t^{\kappa-1}/\theta^{\kappa}$ and κ is estimated to be greater than unity this analysis leads to the conclusion that the hazard function is an increasing function of t.

$\square$

So far only point estimates have been considered. However, interval estimators give an interval and a confidence level for that interval which indicate the precision of the estimator. This is achieved by calculating a standard error for the estimator, which can be done using large sample results with the likelihood function. Theoretical results for the maximum likelihood estimation of the parameters $\boldsymbol{\beta}$ of a model show that in the limit as the sample size n gets large the distribution of the maximum likelihood estimators $\hat{\boldsymbol{\beta}}$ is a normal distribution. If there are r parameters then this limiting (asymptotic) distribution is a multivariate normal distribution. The multivariate normal distribution like the univariate normal distribution defined in Subsection 2.3.4 is defined using the first two population moments and this is done using a vector of expectations $\boldsymbol{\mu}$ and a variance covariance matrix $\mathbf{V}$, which contains the variances in the main diagonal and the covariances in the off-diagonal entries. (The covariance is defined for any pair of random variables as the difference between the expectation of the product of the two random variables and the product of the expectations of these two random variables. The covariance is zero for two *independent* random variables.)

To obtain the variance covariance matrix for $\hat{\beta}$, minus the expectation of the second partial derivatives of the log likelihood function is used to obtain a matrix, $\mathbf{I}_f$, which is known as the *Fisher information* matrix. Under mild regularity conditions the maximum likelihood estimators $\hat{\beta}$ will have a multivariate normal distribution with expectations β and variance covariance matrix given by $\mathbf{I}_f^{-1}$, the inverse of the Fisher information matrix, in the limit as the sample size n increases to infinity. Hence this result can be used for large sample sizes. In practice sample sizes do not have to be very large for this result to give a good approximation providing the proportion of censored observations is small. To use this result it is necessary to evaluate the expectations of the second derivatives of the log likelihood function. Fortunately in practice it is usually sufficient to use the value of the second derivative evaluated at the values observed in the sample, replacing any unknown parameters by their maximum likelihood estimates. The matrix $\mathbf{I}_o$ containing these second derivatives is known as the *observed information* matrix. The square roots of the entries in the main diagonal of $\mathbf{I}_o^{-1}$, the inverse of the observed information matrix, can then be used to give standard errors of the estimators. Also because the distribution is multivariate normal then a covariance of zero for any pair of estimators implies that these estimators are *independently* distributed. (*Note*: This last result is only true because the distributions are normal. It is **not generally** true that zero covariance is equivalent to statistical independence.)

Example 2.6.5 (continued)
Continuing the Weibull distribution example the second derivatives of the log likelihood are

$$\frac{\partial^2 l}{\partial \theta^2} = -\frac{\kappa(\kappa+1)\sum_{i=1}^{n} t_i^{\kappa}}{\theta^{\kappa+2}} + \frac{d\kappa}{\theta^2}$$

$$\frac{\partial^2 l}{\partial \theta \partial \kappa} = \frac{\sum_{i=1}^{n} t_i^{\kappa}}{\theta^{\kappa+1}} + \frac{\kappa \sum_{i=1}^{n} t_i^{\kappa}\log(t_i/\theta)}{\theta^{\kappa+1}} - \frac{d}{\theta}$$

$$\frac{\partial^2 l}{\partial \kappa^2} = -\frac{d}{\kappa^2} - \frac{\sum_{i=1}^{n} t_i^{\kappa}\log(t_i/\theta)^2}{\theta^{\kappa}} .$$

Hence, on substituting the observed sample values and replacing the parameters with their maximum likelihood estimates, the observed information matrix is

$$\begin{pmatrix} 0.000115 & -0.002726 \\ -0.002726 & 9.038249 \end{pmatrix}.$$

Hence, by inverting this matrix, and taking the square roots of the entries in the main diagonal, the standard errors of $\hat{\kappa}$ and $\hat{\theta}$ are 0.3338 and 93.59, respectively.

With these standard errors it is possible to construct 95% confidence intervals. These are (1.008, 2.317) for κ and (435.8, 802.7) for θ, respectively.

$\square$

2.6.4 BAYESIAN METHODS

The third method of estimation is the *Bayesian* method due to the followers of T. Bayes. This method is philosophically different from the previous two methods as it considers the parameter β as the realization of a random variable B, say. The distribution of this random variable is known as the *prior* distribution. Advocates of Bayesian methods combine the prior distribution with the sample observations to produce the *posterior* distribution by using a procedure which is now referred to as *Bayes' rule*.

Bayes' rule
The probability density function $f_{B|\mathbf{T}}(\beta \mid \mathbf{T})$ of the *posterior* distribution of β obtained from the sample of observations $\mathbf{T}$ and the probability density function $f_B(\beta)$ of the *prior* distribution of B is given by

$$f_{B|\mathbf{T}}(\beta \mid \mathbf{T}) = \frac{f_{\mathbf{T}|B}(\mathbf{T} \mid \beta)f_B(\beta)}{f_{\mathbf{T}}(\mathbf{T})},$$

where

$$f_{\mathbf{T}}(\mathbf{T}) = \int_{-\infty}^{\infty} f_{\mathbf{T}|B}(\mathbf{T} \mid \beta)f_B(\beta)d\beta.$$

Bayes' rule produces a distribution for the parameter β, not a point estimator. Thus probability statements can be made about β using the posterior distribution, usually by constructing Bayesian confidence *(credible)* intervals. If a single value is required to give a point estimator then the *mode* of the distribution is often used, though the expected value could be used.

Example 2.6.6
A random sample of observations $t_1, t_2, ..., t_n$ was obtained from a population of failure times with an exponential distribution with parameter λ, as defined in Subsection 2.3.1. A *prior* distribution for λ is chosen to be the gamma distribution, as defined in Subsection 2.3.1, with *shape* parameter α but with *scale* parameter a rather than λ. This prior distribution will have a mode equal to $(\alpha - 1)/a$.

Applying Bayes' rule to this problem gives a *posterior* distribution λ which is also gamma but with shape parameter $\alpha + n$ and scale parameter $a + n\bar{t}$, where $\bar{t}$ is the sample (arithmetic) mean of the observations. So the effect of the information from the observations is to change the parameters of the *prior* gamma distribution. (However, if **no** observations are taken then there is **no** change from the *prior* distribution.) If the mode of the *posterior* distribution is used to give a single value for λ then the value is

$$\frac{\alpha - 1 + n}{a + n\bar{t}}.$$

This mode takes the value $(\alpha - 1)/a$ when $n = 0$ and approaches $1/\bar{t}$ when n is large, which is the answer obtained for the estimator of λ in both Examples 2.6.1 and 2.6.4. For the exponential distribution the reciprocal of λ is the expected value. So considering the reciprocals, the analysis using Bayes' rule is consistent with the view that as a large amount of information is obtained from the sample the 'best' value for the *population* mean (expected value, $1/\lambda$) is the *sample* mean (arithmetic mean of the sample, $\bar{t}$).

Using the example of failure times for the 20 pressure vessels presented in Subsection 1.4.1 with $\alpha = 2$ and $a = 400$, this gives a mode for the prior distribution of λ equal to 0.0025, which is equivalent to an expectation of 400. The mode of the posterior is equal to 0.00176, which is equivalent to an expectation of 567.2. Hence the estimate of the expectation obtained by the use of the prior distribution, corresponding to a low expectation of 400, is slightly less than 575.3, which is the estimate which would be obtained from the estimate of λ given in Examples 2.6.1 and 2.6.4.

□

2.7 Hypothesis testing and goodness-of-fit tests

Another objective of the reliability study may be to decide, when using a parametric model with parameters β, whether the reliability data, which are the observations collected (in a possibly censored or truncated form) from a random sample, are consistent with the parameters β having the value β_0, say. *Inference* is the science which provides methods for answering such questions. The *classical* method is to use the approach of *hypothesis testing* due to J. Neyman and E.S. Pearson. However, there is also an approach based on the use of *likelihood*. The parameter β will be considered as a single variable for ease of explanation. The ideas may be generalized to more variables.

2.7.1   HYPOTHESIS TESTS

The simplest form of the hypothesis test is the situation where there is
a choice between two simple hypotheses or two values of β, say β_0 and
β_1. The preferred value of these two, usually denoted by β_0, is known as
the *null* hypothesis while the other value, β_1, is known as the *alternative*
hypothesis. The aim of the test is to decide whether the observed data are
consistent with the null hypothesis (and so should be *accepted*) or whether
the null hypothesis should be *rejected* (and hence the alternative hypothesis
be accepted). In performing the test it is possible to make two types of
error, namely rejecting the null hypothesis when it is true (known as a *Type
I error*) and secondly accepting the null hypothesis when the alternative
hypothesis is true (known as a *Type II error*). The probability of an error
being made is known as the *size*. It would be nice to minimize the sizes of
both these errors but in general it is necessary to trade one off against the
other.

The quantity given by one minus the size of the Type II error is known as
the *power* of the test. So if there is a choice between two tests which have the
same size of the Type I error (called the *significance level of the test*) then
the test with the larger power would be chosen as it will have the smaller
size of the Type II error. So power is a criterion for choosing between
tests and ideally the *most powerful* test would be preferable. Neyman and
Pearson provided a lemma for constructing the most powerful test.

Neyman–Pearson lemma
In testing two simple hypotheses on the basis of a sample of observations
X, say, of size n, say, the test is defined as rejecting the null hypothesis
for one set of observations (known as the *critical region*) and accepting the
null hypothesis otherwise. The Neyman–Pearson lemma gives a method of
constructing the *critical region* for the *most powerful* test for testing the
null hypothesis against the *alternative* hypothesis amongst all tests with a
given *significance level* and *sample size*.

The simple hypotheses situation is obviously a very idealized situation.
However, it is possible to extend the ideas of hypothesis testing to *composite*
hypotheses, which are defined by using intervals, rather than single values,
for the values of the parameter β. The concept of *power* is extended to the
power function, which is the probability that **X** belongs to the *critical region*
(so that the null hypothesis is rejected), given a certain hypothesis is true.
This is a function of the value of the parameter β and so a 'good' test would
have low values when β takes values that define the null hypothesis and
high values when β takes values that define the alternative hypothesis. The
significance level of the test is defined as the maximum of the significance
levels for tests with a null hypothesis defined by values of β which define

the *composite* null hypothesis. However, there are difficulties. It may not be possible to discriminate between two tests (as was done for simple hypotheses) by deciding that one test is 'uniformly more powerful' than the other.

When using a hypothesis test it is usual to quote the significance level and this is commonly referred to as the 'p-value'. It is conventional to reject the null hypothesis if this p-value is less than 0.05.

Example 2.7.1

Consider carrying out a hypothesis test of a particular value of $\lambda = \lambda_0$, say, for the exponential parametric distribution, given in Subsection 2.3.1, against an alternative hypothesis that $\lambda = \lambda_1$, where $\lambda_0 > \lambda_1$. In Example 2.6.4 it was shown that

$$\prod_{i=1}^{n} f_T(t_i; \lambda) = \lambda^n \exp(-n\lambda/\hat{\lambda}),$$

where $\hat{\lambda} = 1/\bar{t}$ is the maximum likelihood estimator of λ. The Neyman–Pearson lemma defines the critical region as

$$\frac{\prod_{i=1}^{n} f_T(t_i; \lambda_0)}{\prod_{i=1}^{n} f_T(t_i; \lambda_1)} = \frac{\lambda_0^n \exp(-n\lambda_0/\hat{\lambda})}{\lambda_1^n \exp(-n\lambda_1/\hat{\lambda})} < C,$$

which rearranges to give

$$\hat{\lambda} < \frac{n(\lambda_0 - \lambda_1)}{\log((\lambda_0/\lambda_1)^n C)},$$

for a suitable value of C. The value of C is determined by the size of the Type I error, conventionally chosen as 0.05, by using the distribution of $\hat{\lambda}$. Under the null that $\lambda = \lambda_0$ the distribution of $2n\lambda_0/\hat{\lambda}$ is chi-square with $2n$ degrees of freedom (χ^2_{2n}).

Using the example of failure times for the 20 pressure vessels presented in Subsection 1.4.1, consider testing the null hypothesis $\lambda_0 = 0.0025$. Then, using the chi-square distribution with 40 degrees of freedom (χ^2_{40}), the critical region is defined by

$$\hat{\lambda} < 0.00182.$$

Hence as the estimate of λ is 0.00174 the null hypothesis is rejected in favour of the alternative hypothesis that $\lambda < 0.0025$.

$\square$

One popular way to test a hypothesis that $\beta = \beta_0$, say, is to use a Wald test. This simply uses the estimator $\hat{\beta}$ of β and its standard error $se(\hat{\beta})$ and the result that for large sample sizes the ratio $(\hat{\beta} - \beta_0)/se(\hat{\beta})$ will approximately have a standard normal distribution, which has an expected value of 0 and a standard deviation of 1. Hence $(\hat{\beta} - \beta_0)^2/(se(\hat{\beta}))^2$ has a chi-square distribution with 1 degree of freedom (χ_1^2). A hypothesis test based on this statistic is known as a Wald test.

Example 2.7.2

Consider carrying out a hypothesis test of a particular value of $\kappa = 1$ for the Weibull distribution in Example 2.6.5, against an alternative hypothesis that $\kappa = 1$. The null hypothesis is equivalent to assuming the exponential parametric distribution. The Wald statistic is $(1.662 - 1)^2/0.3338^2 = 3.93$ with a p-value of 0.047. The conventional interpretation of this result would be to reject the null hypothesis of $\kappa = 1$, which is consistent with the fact that the confidence interval obtained in Example 2.6.5 does not contain the value 1.

□

2.7.2 LIKELIHOOD RATIO TESTS

In the case of simple hypotheses the Neyman–Pearson lemma leads to considering the ratio of likelihoods to construct the most powerful test. This idea can be extended to composite hypotheses by considering the ratio of the maxima of likelihoods over the respective composite hypotheses. This procedure produces statistics for tests which are known as *likelihood ratio tests*. The procedure often enables a test to be found either easily by tractable mathematical methods or by numerical methods. The behaviour of such tests for large sample sizes has been extensively studied and is well known in many situations.

Example 2.7.3

Consider carrying out a hypothesis test of a particular value of $\lambda = \lambda_0$, say, for the exponential parametric distribution, given in Subsection 2.3.1, against an alternative hypothesis that $\lambda = \lambda_0$. In Example 2.6.4 it was shown that

$$L(\lambda) = \lambda^n \exp(-n\lambda/\hat{\lambda}),$$

where $\hat{\lambda} = 1/\bar{t}$ is the maximum likelihood estimator of λ. Hence the log likelihood function is given by

$$l(\lambda) = n \log \lambda - n\lambda/\hat{\lambda}.$$

Then the test is based on Λ, the likelihood ratio test statistic, which is minus twice the difference of the log likelihoods maximized over the null hypothesis and the alternative composite hypothesis, respectively. Hence as $l(\lambda)$ is maximized at $\lambda = \lambda_0$ and $\lambda = \hat{\lambda}$, respectively, then

$$\Lambda = -2(l(\lambda_0) - l(\hat{\lambda})) = 2n(\log(\hat{\lambda}/\lambda_0) + \lambda_0/\hat{\lambda} - 1),$$

which under the null hypothesis that $\lambda = \lambda_0$ has the chi-square distribution with 1 degree of freedom (χ_1^2), for large sample size, n. Choosing the size of the Type I error, conventionally chosen as 0.05, and using the distribution of Λ gives a critical region for rejecting the null hypothesis defined by

$$2n(\log(\hat{\lambda}/\lambda_0) + \lambda_0/\hat{\lambda} - 1) > 3.84,$$

which is equivalent to
$$\hat{\lambda} < C_1 \text{ or } \hat{\lambda} > C_2$$

for suitable values of C_1 and C_2.

Using the example of failure times for the 20 pressure vessels presented in Subsection 1.4.1, consider testing the null hypothesis $\lambda_0 = 0.0025$. Then the critical region is defined by

$$\hat{\lambda} < 0.00166 \text{ or } \hat{\lambda} > 0.00401.$$

Hence as the estimate of λ is 0.00174 the null hypothesis is accepted as opposed to the alternative hypothesis that $\lambda = 0.0025$. Note the difference between the inference obtained in this example and that obtained in Example 2.7.1. This is because the alternative hypotheses considered are *two sided* as λ can be either less than or greater than λ_0. In Example 2.7.1 it was assumed that for the alternative hypothesis λ was less than λ_0.

□

2.7.3   GOODNESS-OF-FIT TESTS

One area of application where hypothesis testing can be usefully applied is to the question of whether a random sample of observations is a 'good fit' for a favoured hypothesized model. In reliability studies this may often be whether failure times come from a Weibull distribution. One test which can be used to answer this question is the *classical chi-square test* due to K. Pearson.

Example 2.7.4
Again using the example of failure times for the 20 pressure vessels presented in Subsection 1.4.1, consider testing the null hypothesis that p, the

probability that the failure time of a pressure vessel is less than 100 hours, is equal to 0.5. The Pearson chi-square statistic is given by

$$X^2 = \sum_i \frac{(O_i - E_i)^2}{E_i} \, ,$$

where O_i is the number observed in category i and E_i the respective expectation. For the pressure vessels example $E_i = 10$ for times less than 100 hours and for times greater than 100 hours. The observed values are 5 and 15, respectively. Hence $X^2 = 5$. Under the null hypothesis this statistic X^2 has a chi-square distribution with 1 degree of freedom (χ_1^2), for large sample sizes. Hence the null hypothesis of $p = 0.5$ would be rejected at the 0.05 level of significance as the critical region corresponds to $X^2 > 3.84$.

□

2.8   Discussion

In this chapter the main models used in the analysis of failure data in reliability studies were introduced. In Section 2.3 the properties of four important parametric families of distributions, the gamma, Weibull, extreme value, and normal distributions, were described. The first two are defined for positive values, though the other two are defined for both positive and negative values of time. This means that though the first two can be used to model failure times, it is more usual to use the other two (the extreme value and normal) to model the logarithm of time.

As was demonstrated in the examples in Chapter 1, Section 1.4, data obtained from reliability studies are often incomplete owing to censoring. The various ways that these data may be obtained were outlined in Section 2.4. Though parametric models for lifetime distributions are important, the use of non-parametric methods is beneficial and these were illustrated in Section 2.5 for right censored data.

In order to use parametric models for lifetime distributions it is necessary to estimate the values of the parameters of the model. A brief account of the classical, likelihood, and Bayesian methods of estimation was presented in Section 2.6, though the various merits of the methods were not discussed, and these methods are illustrated in later chapters in various examples.

The chapter was concluded in Section 2.7 by a brief account of inferential methods for performing statistical tests to be able to assess whether a hypothesized lifetime distribution is consistent with observed failure times.

3
Analysis of lifetimes with covariates

3.1 Introduction

In many reliability studies as well as the information on the component's (or system's) lifetime there is also available other information about the component (system) or its environment. Typically there may be information on the design of the component or the wear the component has suffered. Hence instead of just recording a set of lifetimes there may be associated with each lifetime other variables. These variables are known as covariates. (The term is often used to cover factors, see Section 3.2, as well as measurable variables.) This extra information should yield more understanding about the performance of the component or system. The more information used the better the understanding, hopefully.

Recently there has been considerable interest in the analysis of such data sets in the reliability and related literature. Part of the interest may be due to the availability of the proportional hazards model, suggested by Cox (1972, 1975). This has focused interest in relating both variables and factors to components' lifetimes. However, the proportional hazards model is not the only model available and a number of approaches predate it, such as the use of the accelerated failure time model (Nelson, 1993) and the Weibull regression model (Smith, 1991). All the models and techniques discussed in this chapter are regression-type techniques.

Ansell (1987) suggested that there are four main reasons for the analysis of these types of data within the reliability literature. These are:
(a) to find significant factors (or variables) which affect lifetime;
(b) to remove nuisance variables which distort analysis;
(c) to increase comprehension of the failure model; and
(d) to produce a better prediction of the failure rate.
Examples of each of these uses appear in the reliability literature, see for example Bendell and Wightman (1985), Ansell and Ansell (1987), Dale (1983), Drury *et al.* (1987), and Jardine and Anderson (1984).

In using regression models the aim is to account for the variation in one variable in terms of other variables or factors. In reliability modelling this usually means relating the variation in the lifetimes to other variables or factors. In many regression analyses the lifetimes are assumed to have a normal distribution. In lifetime studies this is rarely the case; more usually exponential or Weibull models are used. This will affect the algorithms used for estimation, though the general approach to, and concerns which arise from, regression modelling are still applicable in lifetime studies.

For illustration suppose the lifetime, T, of a component is a random variable which has an exponential distribution, with an expectation which is possibly related to a set of variables $\mathbf{z} = (z_1, z_2, ..., z_k)$. Then from Chapter 2, Subsection 2.3.1, the probability density function $f_T(t)$ of T is given by

$$f_T(t) = \begin{cases} \lambda(\mathbf{z}; \boldsymbol{\beta}) \exp(-\lambda(\mathbf{z}; \boldsymbol{\beta})t), & \text{for } t \geq 0, \\ 0, & \text{for } t < 0, \end{cases}$$

where λ is a function of $\mathbf{z}$ and $\boldsymbol{\beta}$, which is a vector of unknown parameters. It is necessary to decide on the appropriate function λ and then it will be necessary to estimate the parameters $\boldsymbol{\beta}$. Usually $\log \lambda$ will be taken to be a linear function.

Historically least squares estimation was used for regression models, though it is common to use maximum likelihood estimation. If maximum likelihood estimation is used then a Newton–Raphson numerical procedure can be applied. Properties of the estimators of the parameters may be determined by exact methods or by use of the asymptotic theory approximations, see Chapter 2, Subsection 2.6.3. The maximum likelihood estimate of $\boldsymbol{\beta}$ will have asymptotically a multivariate normal distribution with mean $\boldsymbol{\beta}$ and a variance which is given by the reciprocal of the Fisher information. The appropriateness of these asymptotic results will usually depend on the sample size but may also depend on other variables such as total observation time or even some of the parameters of the model, see Sweeting (1992). Tests on both the parameters and the models can be constructed using this asymptotic theory. The tests will be discussed in the relevant sections. It is also possible to consider 'goodness-of-fit' of the models to the data. Some of the tests of the appropriateness of the model for the data are based on the residuals. A number of residuals can be defined for the lifetime regression models and these are discussed in Sections 3.3, 3.4, and 3.5.

One aspect which differentiates lifetime regression models from other regression models is censoring or truncation of the data. As stated before it is not unusual that upwards of 90% of the data will be censored, see Chapter 1, Subsection 1.4.2. The estimation procedure must be capable of taking account of this censoring.

3.2 Data and design

For the type of analyses discussed in this chapter as well as the times of failures, or times between failures, there will be data on other variables, usually referred to as covariates or concomitant variables. These extra data typically describe the construction of the component (or system) or the condition in which it functions. This is extra information and may account for the lifetime of the component. There may, of course, be other factors which affect lifetimes which have not been measured. The covariates may be controllable, in that they can be selected by an experimenter, or alternatively they may be nuisance factors outside the control of the experimenter.

Example 3.2.1
An experiment may be designed to investigate the best position of a brake on a wheel. The choice of site for the brake might be controlled by the experimenter, but the road conditions under which the experiment may be run might be beyond the control of the experimenter. So one variable would be the site of the brake and another might be the wetness of the road surface. The variables can be measures or factors. A measure is a direct reading, for example a temperature, the amount of wear, etc. A factor is a variable which is discrete, taking only a limited number of values. For example, the position of the brake could be a factor.

When modelling the effect of a factor it is common practice to use dummy variables. For example, if considering the position of the brake there are two possible sites, A and B, then there are two levels to the factor. Then if the brake was sited at position A the associated variable would take the value 0 and if the brake was sited at position B it would take the value 1.

□

Example 3.2.2
Suppose there is interest in the tensile strength of steel rods; then the carbon content might only be recorded as high, medium, or low. The variable describing the amount of carbon would be a factor with three levels, so it is possible to use three dummy variables: one for high carbon content, one for medium carbon content, and one for low carbon content. If a rod had high carbon content then it would have 1 for the first variable and 0 for the other two. These variables are related, with one out of the three being 1 whilst the other two are 0. Hence there are a range of alternative formulations which can be used. Three formulations are given in Table 3.1.

□

Table 3.1. Three possible formulations for the factor representing the content of carbon in steel rods considered in Example 3.2.2

Carbon content	Formulations		
	Three dummy variables	Two dummy variables	
	$d_1\ d_2\ d_3$	$d_1\ d_2$	$d_1\ d_2$
High	1 0 0	1 0	1 0
Medium	0 1 0	0 1	0 1
Low	0 0 1	$-1\,-1$	0 0

Whilst often reliability data come from unplanned situations, there are some cases when the data do arise from laboratory trials or pre-planned experiments. There are several advantages of designed experiments; the two most important are the clarification of the aim of the experiment and the efficiency of collection of the data that can result from design. Design of experiments is a large field in statistics and has in the past been primarily associated with agricultural experiments, though it has been applied successfully in many engineering contexts, see Grove and Davis (1992). Currently the subject is attracting considerable attention through the drive for quality in production arising out of the quality movements, total quality management, and statistical process control. In the context of reliability, however, the subject has not received sufficient attention, though some authors have addressed this issue recently, see Davis (1991). There are a number of papers on design where the underlying distribution is assumed to be Weibull, see Zelen (1959) and Smith (1991).

It should be remembered, though, that unfortunately the majority of reliability data come from unplanned situations with little control on the number of variables collected, the regularity of measurement of variables, and checks on the accuracy of recording. Hence the analyst may encounter situations where there are a large number of variables to choose from to explain the variation in the lifetimes of the components. Selecting the variables to include in an analysis can be an art in itself and will be discussed in Section 3.8 of this chapter.

3.3   Weibull regression model

The material covered in this section should enable readers to consider both the Weibull regression model and other distributional regression models. The algorithms will, of course, differ for other distributions from those given in this section but the general principles will still hold.

It is usual for regression models to describe one or more of the distribution's parameters in terms of the covariates $\mathbf{z}$. The relationship is usually linear, though this is not always the case. The Weibull distribution, see Chapter 2, Subsection 2.3.2, has a reliability function which can be given by

$$R_T(t) = \exp(-\lambda t^\kappa), \text{ for } t > 0,$$

where $\lambda = 1/\theta^\kappa$ and θ is the scale parameter, and κ is the shape parameter. Each of these parameters could be described in terms of the covariates $\mathbf{z}$, though it is more usual to define either the scale or shape parameter in terms of $\mathbf{z}$. For example, if the scale parameter was chosen then a common model would be to have $\lambda(\mathbf{z}; \boldsymbol{\beta}) = \exp(\boldsymbol{\beta}^T \mathbf{z})$, where the number of covariates $k = r$, the number of parameters. Then the reliability function would be

$$R_T(t \mid \mathbf{z}; \boldsymbol{\beta}) = \exp(-\exp(\boldsymbol{\beta}^T \mathbf{z})t^\kappa), \text{ for } t > 0.$$

The probability density function is given by

$$f_T(t \mid \mathbf{z}; \boldsymbol{\beta}) = \kappa \exp(\boldsymbol{\beta}^T \mathbf{z})t^{\kappa-1} \exp(-\exp(\boldsymbol{\beta}^T \mathbf{z})t^\kappa), \text{ for } t > 0.$$

This model is commonly referred to as the Weibull regression model but there are alternatives which have been studied, see Smith (1991), where the shape parameter is dependent also on the covariate.

There are advantages to reparameterizing this model by taking logs, so that the model takes the form of Gumbel's extreme value distribution, see Chapter 2, Subsection 2.3.3. A reason for this is to produce a model more akin to the normal regression model, but also it allows a more natural extension of the model and hence greater flexibility. Define $Y = \log T$ so that the reliability function is given by

$$R_Y(y \mid \mathbf{z}; \boldsymbol{\beta}) = \exp(-\exp(\kappa y + \boldsymbol{\beta}^T \mathbf{z}))$$

so that

$$E(\log T) = -\frac{\gamma}{\kappa} - \frac{\boldsymbol{\beta}^T \mathbf{z}}{\kappa},$$

where γ is Euler's constant. It is usual to estimate the parameters by using the maximum likelihood approach. Suppose that there are n observations some of which may be right censored. Using the d_i notation introduced in Chapter 2, Subsection 2.4.1, then the log likelihood l is given by

$$l(\kappa, \boldsymbol{\beta}; \mathbf{y}, \mathbf{Z}) = \sum_{i=1}^{n} d_i \log f_Y(y_i, \mathbf{z}_i; \boldsymbol{\beta}) + \sum_{i=1}^{n} (1 - d_i) \log R_Y(y_i, \mathbf{z}_i; \boldsymbol{\beta}),$$

where $\mathbf{y} = (y_1, y_2, ..., y_n)^T$, $\mathbf{Z} = (\mathbf{z}_1, \mathbf{z}_2, ..., \mathbf{z}_n)$, and $\mathbf{z}_i = (z_{i1}, z_{i2}, ..., z_{ir})^T$. Substituting for f_Y and R_Y gives

$$l(\kappa, \boldsymbol{\beta}; \mathbf{y}, \mathbf{Z}) = d \log \kappa + \sum_{i=1}^{n} d_i(\kappa y_i + \boldsymbol{\beta}^T \mathbf{z}_i) - \sum_{i=1}^{n} \exp(\kappa y_i + \boldsymbol{\beta}^T \mathbf{z}_i),$$

where $\sum_{i=1}^{n} d_i = d$. The equations for the maximum likelihood estimators of the β and κ are

$$\frac{\partial l}{\partial \beta_j} = \sum_{i=1}^{n} d_i z_{ij} - \sum_{i=1}^{n} z_{ij} \exp(\kappa y_i + \boldsymbol{\beta}^T \mathbf{z}_i) = 0;$$

$$\frac{\partial l}{\partial \kappa} = \frac{d}{\kappa} + \sum_{i=1}^{n} d_i y_i - \sum_{i=1}^{n} y_i \exp(\kappa y_i + \boldsymbol{\beta}^T \mathbf{z}_i) = 0.$$

The second derivatives of the log likelihood are

$$\frac{\partial^2 l}{\partial \beta_j^2} = -\sum_{i=1}^{n} z_{ij}^2 \exp(\kappa y_i + \boldsymbol{\beta}^T \mathbf{z}_i),$$

$$\frac{\partial^2 l}{\partial \beta_j \beta_k} = -\sum_{i=1}^{n} z_{ij} z_{ik} \exp(\kappa y_i + \boldsymbol{\beta}^T \mathbf{z}_i),$$

$$\frac{\partial^2 l}{\partial \beta_j \partial \kappa} = -\sum_{i=1}^{n} z_{ij} y_i \exp(\kappa y_i + \boldsymbol{\beta}^T \mathbf{z}_i),$$

$$\frac{\partial^2 l}{\partial \kappa^2} = -\frac{d}{\kappa^2} - \sum_{i=1}^{n} y_i^2 \exp(\kappa y_i + \boldsymbol{\beta}^T \mathbf{z}_i).$$

Solutions to these equations for the maximum likelihood estimators can be obtained using an iterative Newton–Raphson procedure. However, Aitkin and Clayton (1980) suggest that convergence is slow for κ using such a procedure. They suggest fitting the Weibull regression model using the statistical software package GLIM. The procedure consists of iteratively fitting a Poisson distribution until convergence is reached. The shape parameter (κ) is usually set initially to 1 and is updated by using fitted values of the Poisson mean, see Aitkin and Clayton (1980) and Roger (1985).

Table 3.2. Lifetimes (in cycles) of sodium sulphur batteries

Batch 1	164	164	218	230	263	467	538	639	669
	917	1148	1678+	1678+	1678+	1678+			
Batch 2	76	82	210	315	385	412	491	504	522
	646+	678	775	884	1131	1446	1824	1827	2248
	2385	3077							

Note: Lifetimes with + are right censored observations, not failures.

Standard errors may be obtained by use of the second derivatives to obtain the observed information matrix, $\mathbf{I}_o$, as was explained in Chapter 2, Subsection 2.6.3, and this matrix is usually calculated in the standard statistical software packages. The estimated variance covariance matrix will be $\mathbf{I}_o^{-1}$ and the standard errors will be the square roots of the diagonal elements of this inverted matrix.

Example 3.3.1
Ansell and Ansell (1987) analysed the data given in Table 3.2 in a study of the performance of sodium sulphur batteries. The data consist of lifetimes (in cycles) of two batches of batteries. The covariate vector for the ith battery is given by $\mathbf{z}_i = (z_{i1}, z_{i2})^T$, where $z_{i1} = 1$ and z_{i2} represents whether the battery comes from batch 1 or batch 2, so that

$$z_{i2} = \begin{cases} 0, & \text{if battery } i \text{ is from batch 1,} \\ 1, & \text{if battery } i \text{ is from batch 2.} \end{cases}$$

Hence β_2 represents the difference in performance between batteries from batch 2 and batch 1.

Fitting the Weibull regression model results in $\hat{\beta}_2 = 0.0156$ and $\hat{\kappa} = 1.127$. Using the observed information matrix the standard error of $\hat{\beta}_2 = 0.386$. Hence a 95% confidence interval for β_2 is $(-0.740, 0.771)$.

An obvious test to perform is to see if β_2 is non-zero. If it is non-zero this would imply there is a difference between the two batches of batteries. The hypotheses are:

$$H_0 : \beta_2 = 0.$$
$$H_1 : \beta_2 \neq 0.$$

The log likelihood evaluated under H_0 is -49.7347 and under H_1 is -49.7339. Hence the likelihood ratio test statistic has a value of 0.0016. Under the null hypothesis the test statistic has a χ^2 distribution with 1 degree of freedom. Therefore the hypothesis that β_2 is zero, which is equivalent to no difference between the batches, is accepted. There are alternative tests, see Lawless (1982). Using the estimate $\hat{\beta}_2$ of β_2 and its standard error gives a Wald statistic of 0.0016, almost the same value as the likelihood ratio test statistic, and hence leads to the same inference. Both these inferences are consistent with the confidence interval, as it includes zero.

□

Examination of the goodness-of-fit and the appropriateness of the assumptions made in fitting the regression model can be based on graphical approaches using residuals, see Smith (1991). The Cox and Snell generalized residuals, see Cox and Snell (1968), are defined as

$$e_i = -\log R_T(t_i \mid \mathbf{z}_i; \hat{\boldsymbol{\beta}}),$$

where $R_T(t_i \mid \mathbf{z}_i; \hat{\boldsymbol{\beta}})$ is the reliability function evaluated at t_i and $\mathbf{z}_i$ with estimates $\hat{\boldsymbol{\beta}}$.

Cox and Snell (1968) provided a first-order correction to these residuals, though in many cases these are not used, see Smith (1991). Obviously one problem that arises is with the residuals for censored observations and authors generally, see Lawless (1982), suggest using

$$e_i = -\log R_T(t_i \mid \mathbf{z}_i; \hat{\boldsymbol{\beta}}) + 1.$$

Cox and Snell residuals should be independent and identically distributed (IID) random variables with a unit exponential distribution, i.e. with an expectation of one. From the result given in Chapter 1, Example 1.6.3, the cumulative hazard function is a linear function with a slope of one. Also the cumulative hazard function is minus the log of the reliability function. Hence a plot of minus the log of the estimated reliability function for the residuals, $-\log \hat{R}(e_i)$, against e_i should be roughly linear with a slope of one when the model is adequate. Other graphs can be informative; see Smith (1991) where plots against covariates and plots of subgroups of data are considered.

Example 3.3.1 (continued)
Using the data in Example 3.3.1 and fitting the Weibull regression model the generalized residuals have been calculated and are presented in Fig. 3.1. The plot is of minus the log of the reliability function of the generalized residuals against the generalized residuals. Since some of the points are

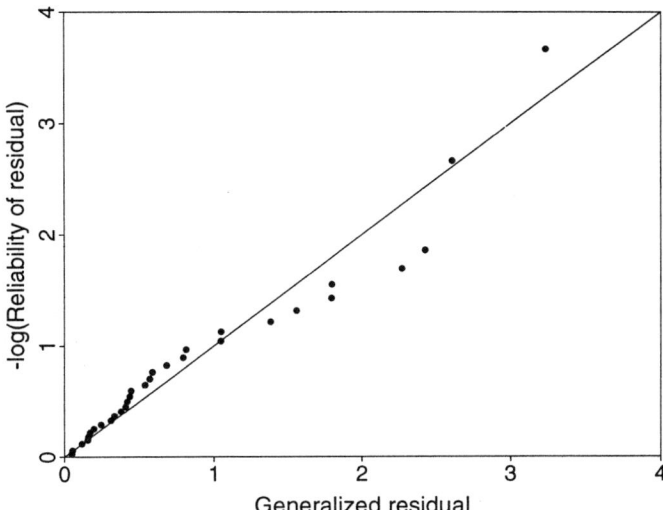

Fig. 3.1. Plot of the generalized residuals of the Weibull regression model for the sodium sulphur battery data

far from the line with slope 1 it would seem that the current model is not necessarily appropriate. Further investigation would be required to clarify if this was due to the choice of the distribution or the current model.

□

The approach described above is applicable to many distributions, though, of course, the algorithms for obtaining the estimates will vary. A special case of the Weibull regression model is the exponential regression model, when the shape parameter is taken to be 1 ($\kappa = 1$). This has been studied by M. Glasser (1967), Cox and Snell (1968), and Lawless (1982). There are a number of other lifetime regression models which have been studied: gamma, log-logistic, and lognormal, see Lawless (1982). Many of these models are covered by the general term location-scale models, or accelerated failure time models. The general model will be explored in Section 3.4.

Selection of an appropriate distribution model depends both on the physical context and on the actual fit achieved. There ought to be good physical justification for fitting a distribution using the context under study. Basing the decision about the distribution purely on fit can be very misleading, especially if there are a number of possible covariates to be chosen.

It is always possible by fitting extra variables to achieve a better fit to a set of data, though such a fit will have little predictive power.

3.4   Accelerated failure time model

The Weibull regression model discussed in the last section can be regarded as an example of an accelerated failure time model. Accelerated failure time models were originally devised to relate the performance of components put through a severe testing regime to a component's more usual lifetime. It was assumed that a variable or factor, such as temperature or number of cycles, could be used to describe the severity of the testing regime. This problem has been considered by a number of authors; see Nelson (1993) for a comprehensive account of this area.

Suppose in a study that the covariate is z, which can take the values 0 and 1, and that it is assumed that the hazard functions are

$$\lambda(t \mid z = 0) = \lambda_0$$

and

$$\lambda(t \mid z = 1) = \phi\lambda_0,$$

so that ϕ is the *relative risk* for $z = 1$ versus $z = 0$.

Then

$$R(t \mid z = 1) = R(\phi t \mid z = 0)$$

and, in particular,

$$E(T \mid z = 1) = \frac{E(T \mid z = 0)}{\phi}.$$

So the time for components with $z = 1$ is passing at a rate ϕ faster than for the components with $z = 0$. Hence the name of the model.

The model can be extended as follows. Suppose ϕ is replaced by $\phi(z)$ with $\phi(0) = 1$; then

$$R(t \mid z) = R(\phi(z)t \mid z = 0),$$

and hence

$$\lambda(t \mid z) = \phi(z)\lambda(\phi(z)t \mid z = 0)$$

and

$$E(T \mid z) = \frac{E(T \mid z = 0)}{\phi(z)}.$$

In using the model for analysis a parametric model is specified for $\phi(z)$ with β as the parameter, which will be denoted by $\phi(z; \beta)$. A typical choice would be

$$\phi(z; \beta) = \exp(\beta z).$$

This choice leads to a linear regression model for $\log T$ as $\exp(\beta z)T$ has a distribution which does not depend on z. Hence $\log T$ is given by

$$\log T = \mu_0 - \beta z + \epsilon,$$

where μ_0 is $E(\log T \mid z = 0)$ and ϵ is a random variable whose distribution does not depend on the covariate z.

To estimate β there is the need to specify the distribution. If the distribution of T is lognormal then least squares estimation may be used as $\mu_0 + \epsilon$ has a normal distribution, see Chapter 2, Subsection 2.3.4. If the distribution of T is Weibull with a shape parameter κ, see Chapter 2, Subsection 2.3.2, then $\kappa(\mu_0 + \epsilon)$ has a standard extreme value distribution, see Chapter 2, Subsection 2.3.3. Hence, as was stated at the beginning of this section, an example of the accelerated failure time model is the Weibull regression model studied in Section 3.3. Other such models have been widely applied in reliability, see Cox (1964), Fiegl and Zelen (1965), Nelson and Hahn (1972), Kalbfleisch (1974), Farewell and Prentice (1977), and Nelson (1993). However, they are not regularly applied by reliability engineers because of the perceived difficulties with estimation because the estimators vary depending on the distribution. There are a number of GLIM macros for specific distributions and the general approach of Aitkin and Clayton (1980) encompasses a number of these distributions. These models can also be fitted using the statistical software package SAS with PROC LIFEREG. Plotting techniques, such as using Cox and Snell generalized residuals as defined for the Weibull regression model, may be used for assessing the appropriateness of the model.

Example 3.4.1
Elsayed and Chan (1990) presented data collected from tests for the time-dependent dielectric breakdown of metal–oxide–semiconductor integrated circuits, which was described in Chapter 1, Subsection 1.4.4, with the data given in Table 1.6. The data consist of times to failure (in hours) for three different temperatures (170 °C, 200 °C, and 250 °C). Elsayed and Chan (1990) suggest a model where the covariate of interest is the inverse of the absolute temperature. So the covariate vector for the ith circuit is given by $\mathbf{z}_i = (z_{i1}, z_{i2})^T$, where $z_{i1} = 1$ and z_{i2} represents the inverse absolute temperature at which the test was performed, and takes the three values 0.001911, 0.002113, and 0.002256. Hence β_2 represents the coefficient of the inverse absolute temperature covariate.

Fitting the Weibull regression model results in $\hat{\beta}_2 = -7132.4$ and $\hat{\kappa} = 0.551$. Using the observed information matrix the standard error of $\hat{\beta}_2$ is 1222.0. Hence a 95% confidence interval for β_2 is $(-9527.5, -4737.1)$. This interval indicates that β_2 is non-zero and this would imply there is a difference in the performance of the circuits at the different temperatures. This can be confirmed by performing a hypothesis test to see if β_2 is non-zero. The hypotheses are

$$H_0 : \beta_2 = 0.$$
$$H_1 : \beta_2 = 0.$$

The log likelihood evaluated under H_0 is -148.614 and under H_1 is -130.112. Hence the likelihood ratio test statistic has a value of 37.00. Under the null hypothesis the test statistic has a χ^2 distribution with 1 degree of freedom. Therefore the hypothesis that β_2 is zero is not accepted and this implies there is a difference between the circuits depending on the temperatures of the tests. Using the estimate $\hat{\beta}_2$ of β_2 and its standard error gives a Wald statistic of 34.06, almost the same value as the likelihood ratio test statistic, and hence leads to the same inference.

The generalized residuals are presented in Fig. 3.2. The plot is of minus the log of the reliability function of the generalized residuals against the generalized residuals. The points lie closer to a line with slope 1 than was the case in Fig. 3.1 for Example 3.3.1. So the current model may be appropriate.

$\square$

3.5 Proportional hazards model

This model has been widely used in reliability studies by a number of authors: Bendell and Wightman (1985), Ansell and Ansell (1987), and Jardine and Anderson (1984). A diversity of uses have been found for the model.

The model Cox (1972) proposed assumed that the hazard function for a component could be decomposed into a baseline hazard function and a function dependent on the covariates. The hazard function at time t with covariates $\mathbf{z}$, $\lambda(t, \mathbf{z})$ would be expressed as

$$\lambda(t \mid \mathbf{z}) = \psi[\lambda_0(t), \phi(\mathbf{z}; \boldsymbol{\beta})],$$

where ψ would be an arbitrary function, $\lambda_0(t)$ would be the baseline hazard function, ϕ would be another arbitrary function of the covariates, $\mathbf{z}$, and

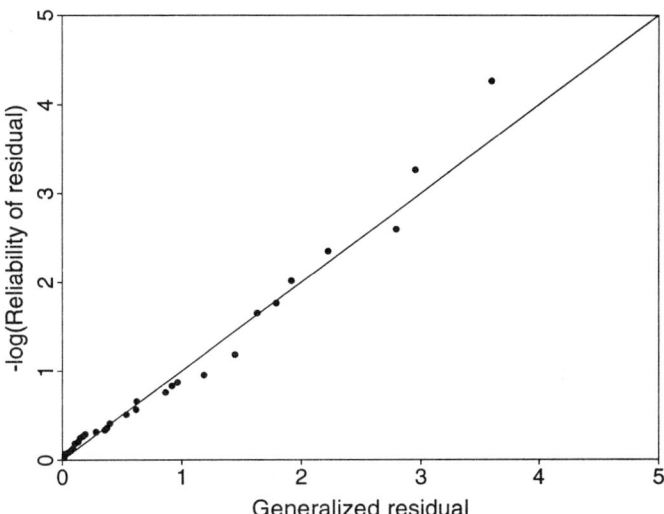

Fig. 3.2. Plot of the generalized residuals of the Weibull regression model for the semiconductor integrated circuit data

β the parameters of the function ϕ. Cox (1972) suggested that ψ might be a multiplicative function and that ϕ should be the exponential with a linear predictor for the argument. This *proportional hazards* model has the advantage of being well defined for most purposes, given that a hazard function must be non-negative. It yields

$$\lambda(t \mid \mathbf{z}) = \lambda_0(t) \exp\left(\boldsymbol{\beta}^T \mathbf{z}\right).$$

However, this is only one possible selection for ψ and ϕ. Using the multiplicative formulation it is usual to define $\phi(\mathbf{z}; \boldsymbol{\beta})$ so that $\phi(\mathbf{0}; \boldsymbol{\beta}) = 1$ so that $\phi(\mathbf{z}; \boldsymbol{\beta})$ is the relative risk for a component with covariate $\mathbf{z}$ compared with a component with covariate $\mathbf{z} = \mathbf{0}$. Thus the reliability function is given by

$$R(t \mid \mathbf{z}) = R(t \mid \mathbf{z} = \mathbf{0})^{\phi(\mathbf{z}, \boldsymbol{\beta})},$$

where $R(t \mid \mathbf{z} = \mathbf{0})$, often denoted simply as $R_0(t)$, is the baseline reliability function. Etezardi-Amoli and Ciampi (1987), amongst others, have considered an alternative additive model.

Cox (1972) considered the case in which the hazard function is a semi-parametric model; $\lambda_0(t)$ is modelled non-parametrically (or more accurately using infinitely many parameters). It is possible to select a specific parametric form for $\lambda_0(t)$, which could be a hazard function from one of the families of distributions discussed in Chapter 2, Section 2.3.

In the case of the semi-parametric model Cox (1975) introduced the concept of *partial* likelihood to tackle the problems of statistical inference. This has been further supported by the work of Andersen and Gill (1982).

In Cox's approach the partial likelihood function is formed by considering the components at risk at each of the n_0 failure times $t_{[1]}, t_{[2]}, t_{[3]}, \ldots, t_{[n_0]}$, as defined in Chapter 2, Section 2.5. This produces a function which does not depend on the underlying distribution and can therefore be used to obtain estimates of $\boldsymbol{\beta}$. The partial likelihood function $L(\boldsymbol{\beta})$ is given by

$$L(\boldsymbol{\beta}) = \prod_{i=1}^{n_0} \frac{\exp(\boldsymbol{\beta}^T \mathbf{z}_{[i]})}{\sum_{w \in R_i} \exp(\boldsymbol{\beta}^T \mathbf{z}_w)} ,$$

where R_i is the risk set and $\mathbf{z}_i$ is the observed covariate for the failure at time t_i. This can be maximized directly using the Newton–Raphson procedure or by use of an E–M algorithm.

Ties often occur in practice in data and adjustments should be made to the estimation procedure to account for ties. Breslow (1974) and Peto (1972) have made suggestions for replacing the partial likelihood function by

$$L(\boldsymbol{\beta}) = \prod_{i=1}^{n_0} \frac{\exp(\boldsymbol{\beta}^T \mathbf{S}_{[i]})}{(\sum_{w \in R_i} \exp(\boldsymbol{\beta}^T \mathbf{z}_w))^{m_i}} ,$$

where m_i is the number of failure times equal to $t_{[i]}$ and $\mathbf{S}_{[i]} = (S_{[i]1}, S_{[i]2}, \ldots, S_{[i]r})$ is the sum of the covariate vectors $\mathbf{z}_i$ for this time.

The first differentials of the log of the partial likelihood function $l(\boldsymbol{\beta})$ are

$$\frac{\partial l(\boldsymbol{\beta})}{\partial \beta_j} = \sum_{i=1}^{n_0} \left(S_{[i]j} - \frac{m_i \sum_{w \in R_i} z_{wj} \exp(\boldsymbol{\beta}^T \mathbf{z}_w)}{\sum_{w \in R_i} \exp(\boldsymbol{\beta}^T \mathbf{z}_w)} \right) ,$$

for $j = 1, \ldots, r$. These derivatives can be equated to zero and solved to give estimators $\hat{\boldsymbol{\beta}}$ of $\boldsymbol{\beta}$.

The second differentials are

$$\frac{\partial^2 l(\boldsymbol{\beta})}{\partial \beta_j \partial \beta_k} = \sum_{i=1}^{n_0} m_i \left[\frac{\sum_{w \in R_i} z_{wj} z_{wk} \exp(\boldsymbol{\beta}^T \mathbf{z}_w)}{\sum_{w \in R_i} \exp(\boldsymbol{\beta}^T \mathbf{z}_w)} - \frac{(\sum_{w \in R_i} z_{wj} \exp(\boldsymbol{\beta}^T \mathbf{z}_w))(\sum_{w \in R_i} z_{wk} \exp(\boldsymbol{\beta}^T \mathbf{z}_w))}{\left(\sum_{w \in R_i} \exp(\boldsymbol{\beta}^T \mathbf{z}_w) \right)^2} \right] .$$

These second derivatives can be evaluated at $\hat{\boldsymbol{\beta}}$ to produce minus an 'information' matrix, whose inverse can be used to obtain standard errors.

Example 3.5.1

Returning to the sodium sulphur battery data used in Example 3.3.1, put $z = z_2$, which was defined in Example 3.3.1 to indicate whether the battery comes from batch 1 or batch 2. (*Note:* The variable $z_1 = 1$ is not required as the arbitrary term $\lambda_0(t)$ will contain any arbitrary constant which was provided in the Weibull regression model in Example 3.3.1 by the parameter β_1.) Then $\hat{\beta} = -0.0888$. Using the information matrix, the standard error of $\hat{\beta}$ is 0.4034. Hence a 95% confidence interval for β is $(-0.879, 0.702)$.

A test can be performed to see if β is non-zero. If it is non-zero this would imply there is a difference between the two batches of batteries. The hypotheses are

$$
\begin{aligned}
H_0 : \beta &= 0. \\
H_1 : \beta &= 0.
\end{aligned}
$$

It is possible to use the partial log likelihood, which when evaluated under H_0 is -81.262 and under H_1 is -81.238. Hence the 'likelihood' ratio test statistic (twice the difference between these partial log likelihoods) has a value of 0.048. Under the null hypothesis this test statistic can be shown to have a χ^2 distribution with 1 degree of freedom, in the same way as for likelihood. Therefore the hypothesis that β is zero is accepted and this implies there is no difference between the batches.

□

An alternative non-parametric approach can be taken in the case when comparing two distributions to see if they are the same. In the case of two groups a test of whether $\beta = 0$ is equivalent to testing whether the two reliability functions, $R_1(t)$ for group 1 and $R_2(t)$ for group 1, are the same. The hypotheses are

$$
\begin{aligned}
H_0 : R_2(t) &= R_1(t) \\
H_1 : R_2(t) &= R_1(t)^{\exp(\beta)}, \text{ for } \beta \text{ not equal to 0.}
\end{aligned}
$$

Suppose $t_{[i]}, i = 1, ..., n_0$, are the ordered failure times and r_{1i} and r_{2i} are the number at risk at time $t_{[i]}$ for group 1 and group 2, respectively, and let m_{1i} be the number of failures of group 1 and let m_{2i} be the number of failures of group 2 at time $t_{[i]}$; then a test statistic would be U^2/I, where

$$
U = \sum_{i=1}^{n_0} \left(m_{2i} - \frac{m_i r_{2i}}{r_i} \right),
$$

$$I = \sum_{i=1}^{n_0} \frac{m_i(r_i - m_i)r_{1i}r_{2i}}{r_i^2(r_i - 1)} ,$$

$m_i = m_{1i} + m_{2i}$, and $r_i = r_{1i} + r_{2i}$. Under the null hypothesis this test statistic can be shown to have a χ^2 distribution with 1 degree of freedom. This procedure was originally proposed by Mantel (1966).

Example 3.5.2

Returning to Example 3.5.1, a test of whether $\beta = 0$ is equivalent to testing whether the reliability functions are the same for both batches of batteries. Now $U = 0.542$ and $I = 6.052$ and hence $U^2/I = 0.0485$. This statistic, though not identical, is similar to that obtained from the likelihood ratio in Example 3.5.1, and hence leads to the same conclusion. Therefore the null hypothesis that β is zero is accepted.

□

These examples have used a categorical covariate which indicates different groups of observations. The next example uses a continuous covariate.

Example 3.5.3

Returning to the semiconductor integrated circuit data used in Example 3.4.1, put $z = z_2$, which was defined in Example 3.4.1 as the inverse of the absolute temperature. (*Note*: Again the variable $z_1 = 1$ is not required as the arbitrary term $\lambda_0(t)$ will contain any arbitrary constant which was provided in the Weibull regression model in Example 3.4.1 by the parameter β_1.) Then $\hat{\beta} = -7315.0$. Using the information matrix, the standard error of $\hat{\beta}$ is 1345.0. Hence a 95% confidence interval for β is $(-9951.2, -4678.8)$.

This interval indicates that β is non-zero and this would imply there is a difference in the performance of the circuits at the different temperatures. This can be confirmed by performing a test to see if β is non-zero. The hypotheses are

$$H_0 : \beta = 0.$$
$$H_1 : \beta = 0.$$

It is possible to use the partial log likelihood, which when evaluated under H_0 is -190.15 and under H_1 is -173.50. Hence the 'likelihood' ratio test statistic (twice the difference between these partial log likelihoods) has a value of 33.31. Under the null hypothesis this test statistic can be shown to have a χ^2 distribution with 1 degree of freedom, in the same way as for likelihood. Therefore the hypothesis that β is zero is not accepted and this implies there is a difference between the circuits depending on the temperatures of the tests.

Using the estimate $\hat{\beta}$ of β and its standard error gives a Wald statistic of 29.60, almost the same value as the likelihood ratio test statistic, and hence leads to the same inference.

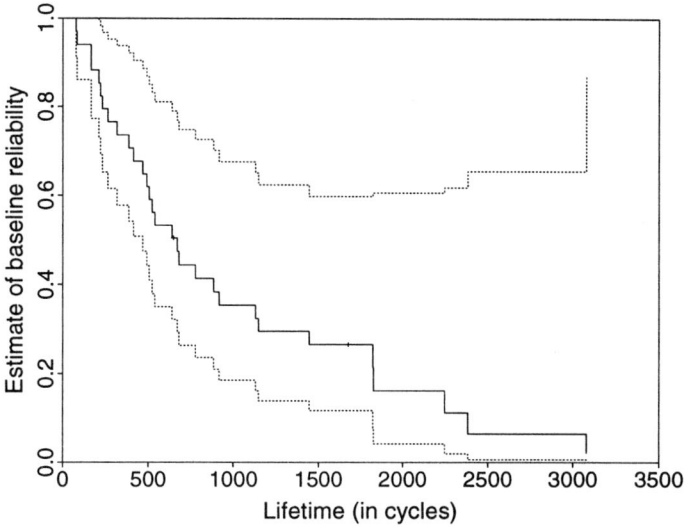

Fig. 3.3. Plot of the baseline reliability function for the proportional hazards model for the sodium sulphur battery data with 95% confidence limits

So far the estimation of β has been considered so that the relative risk can be estimated.

□

3.5.1 ESTIMATION OF THE BASELINE RELIABILITY FUNCTION

The reliability function has to be estimated and the usual approach is to estimate first the baseline reliability function, $R_0(t)$. This reliability function is first described as a product of the α so that

$$R_0(t) = \prod_{i:t_{[i]}<t} \alpha_i,$$

where $t_{[i]}$ are the ordered failure times. Hence the baseline reliability function is estimated at the failure times of the data. It is possible to derive a likelihood for the α and to obtain equations for the estimators $\hat{\alpha}$, which if there is only a single failure at $t_{[i]}$ are given by

$$\hat{\alpha}_i^{\exp(\hat{\beta}^T \mathbf{z}_{[i]})} = 1 - \frac{\exp(\hat{\beta}^T \mathbf{z}_{[i]})}{\sum_{w \in R_i} \exp(\hat{\beta}^T \mathbf{z}_w)},$$

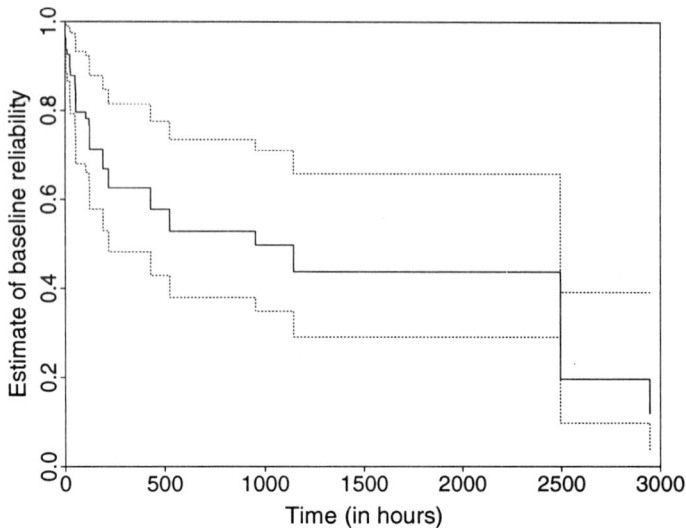

Fig. 3.4. Plot of the baseline reliability function for the proportional hazards model for the semiconductor integrated circuit data with 95% confidence limits

but otherwise $\hat{\alpha}$ has to be found by an iterative method. A suggested initial approximation, see Lawless (1982), is

$$\hat{\alpha}_i = \exp\left(\frac{-m_i}{\sum_{w \in R_i} \exp(\hat{\boldsymbol{\beta}}^T \mathbf{z}_w)}\right),$$

which can be used to obtain the estimate of $R_0(t)$ and hence of $R(t)$.

Example 3.5.4

The estimate of the baseline reliability function for the data on sodium sulphur batteries introduced in Example 3.3.1 with the estimate of $\hat{\beta}$ of β as given in Example 3.5.1 are given in Fig. 3.3 with 95% confidence limits.

□

Baseline has been defined as the case with covariate $z = 0$. However, this is not always a sensible choice of the covariate. This is illustrated in the next example.

Example 3.5.5

For the semiconductor integrated circuit data used in Example 3.4.1, $\mathbf{z} = z$ was defined as the inverse of the absolute temperature. This will only be

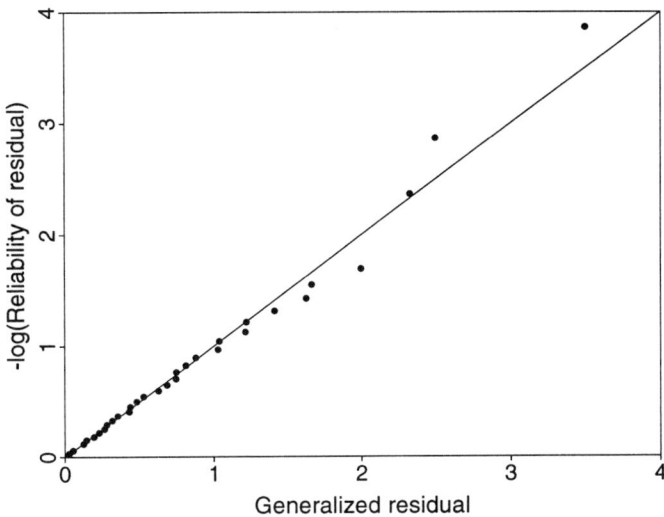

Fig. 3.5. Plot of the generalized residuals of the proportional hazards model for the sodium sulphur battery data

zero if the temperature is infinitely large. Hence it makes more sense to take the baseline value to be one of the temperatures used in the tests. The *smallest* temperature will be chosen, which corresponds to 0.002256, the *largest* value of the inverse absolute temperature. Then the estimate of the baseline reliability function for the data with the estimate of β as given in Example 3.5.3 are given in Fig. 3.4 with 95% confidence limits.

□

3.5.2 RESIDUAL PLOTS

Given the estimates of the reliability function it is then possible to obtain the Cox and Snell generalized residuals which are for an uncensored time

$$e_i = [-\log \hat{R}_0(t_i)] \exp(\hat{\boldsymbol{\beta}}^T \mathbf{z}_i)$$

and for a right censored lifetime

$$e_i = [-\log \hat{R}_0(t_i)] \exp(\hat{\boldsymbol{\beta}}^T \mathbf{z}_i) + 1.$$

These residuals have approximately a unit exponential distribution, i.e. with an expectation of one. From the result given in Chapter 1, Example

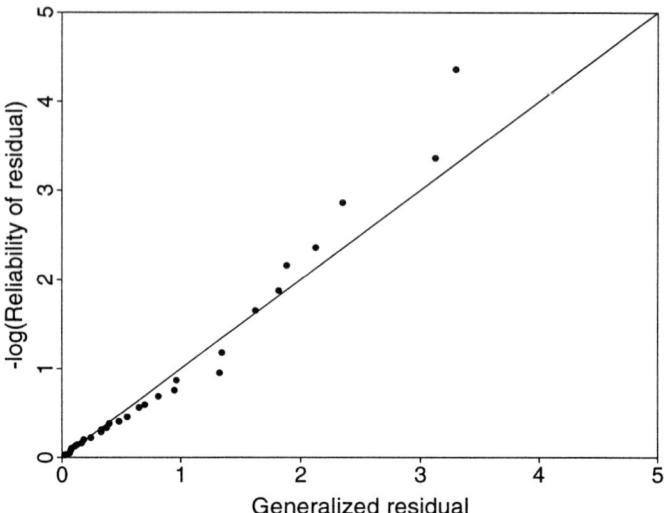

Fig. 3.6. Plot of the generalized residuals of the proportional hazards model for the semiconductor integrated circuit data

1.6.3, the cumulative hazard function is a linear function with a slope of one. Also the cumulative hazard function is minus the log of the reliability function. Hence a plot of minus the log of the reliability function of the residuals against their value should be approximately a straight line with a slope of one. Any departure from such would be a cause for concern.

Example 3.5.6
Using the sodium sulphur battery data from Example 3.3.1 the generalized residuals have been calculated and a plot of minus the log of their reliability function against the generalized residuals is given in Fig. 3.5. There seems to be a better agreement with the straight line than that obtained before in Fig. 3.1 with the Weibull regression model.

□

Example 3.5.7
Using the semiconductor integrated circuit data considered in Example 3.4.1 the generalized residuals have been calculated and a plot of minus the log of their reliability function against the generalized residuals is given in Fig. 3.6. There seems be no better agreement with the straight line than that obtained before in Fig. 3.2 with the Weibull regression model.

□

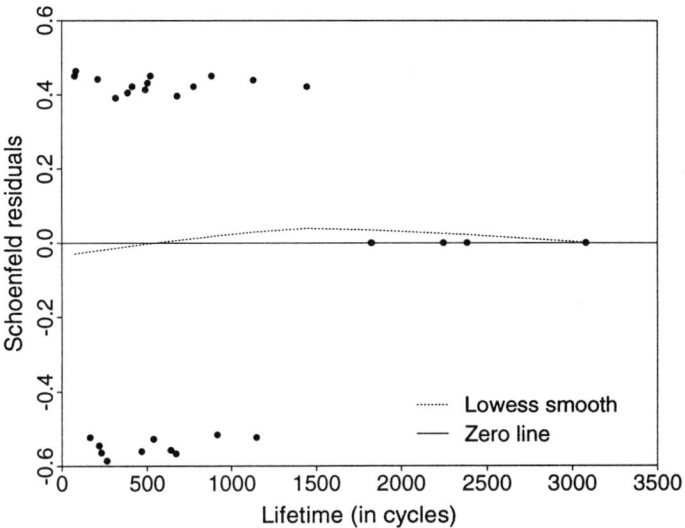

Fig. 3.7. Plot of the Schoenfeld residuals for batch of the proportional hazards model for the sodium sulphur battery data

Other residuals have been suggested for the Cox model. Schoenfeld (1982) suggested partial residuals to examine the proportionality assumption made in the Cox model. These are defined for the jth covariate as follows:

$$r_{ij} = z_{ij} - E(z_{ij} \mid R_i),$$

where $E(z_{ij} \mid R_i)$ is given by

$$E(z_{ij} \mid R_i) = \frac{\sum_{w \in R_i} z_{wj} \exp(\hat{\boldsymbol{\beta}}^T \mathbf{z}_w)}{\sum_{w \in R_i} \exp(\hat{\boldsymbol{\beta}}^T \mathbf{z}_w)} \, .$$

These residuals can then be plotted against time, and if the proportional hazards assumption holds then the residuals should be randomly scattered about zero, with no time trend.

Example 3.5.8
Using the sodium sulphur battery data from Example 3.3.1 and fitting the proportional hazards model, the Schoenfeld residuals are calculated and are presented in Fig. 3.7 as a plot of the residuals against the time. A non-parametric estimate of the regression line, the 'lowess' line, see Cleveland

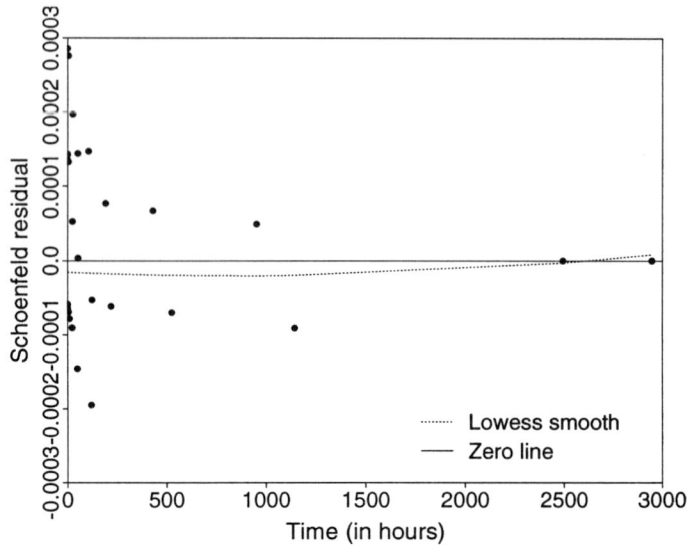

Fig. 3.8. Plot of the Schoenfeld residuals for the inverse of absolute temperature of the proportional hazards model for the semiconductor integrated circuit
data

(1979), is included on the plot as well as the zero residual line. There seems
to be no significant trend. A test for *linear* trend can be used. For this
test the statistic is 0.0654, which would be from a χ^2 distribution with 1
degree of freedom if there was no linear trend. Hence there is no evidence
of a linear trend and it is probably safe to accept the proportional hazards
assumption.

□

Example 3.5.9
Using the semiconductor integrated circuit data as in Example 3.4.1 and
fitting the proportional hazards model, the Schoenfeld residuals are calculated and are presented in Fig. 3.8 as a plot of the residuals against the
time. A non-parametric estimate of the regression line, the 'lowess' line,
is included on the plot as well as the zero residual line. For the test for
linear trend the statistic is 1.60, which would be from a χ^2 distribution
with 1 degree of freedom if there was no linear trend. Hence there is no
evidence of a linear trend. However, it is not clear whether to accept the
proportional hazards assumption.

□

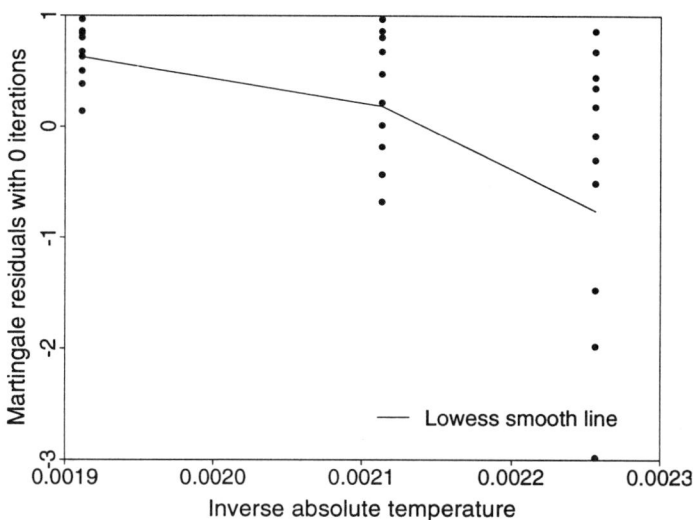

Fig. 3.9. Plot of the martingale residuals versus the inverse of the absolute temperature of the null proportional hazards model for the semiconductor integrated circuit data

A number of statistical software packages facilitate proportional hazards modelling, including SAS and S-PLUS. There are also GLIM macros for proportional hazards models, based on work by Whitehead (1980).

Therneau *et al.* (1990) suggest two alternative residuals, a martingale residual and a deviance residual. Except in the case of discrete covariates these residuals are far from simple to calculate; however, statistical software is available for their estimation, for example the SAS procedure PROC PHREG and S-PLUS functions **coxreg** and **agreg**. In the case of discrete covariates the martingale residuals are a transformation of the Cox and Snell generalized residuals. The deviance residuals are a transformation of the martingale residuals to correct for skewness. Therneau *et al.* (1990) suggest that the martingale residuals are useful in deciding about (a) appropriate functional relationships between the covariates and their survival, (b) proportionality of the hazard functions, and (c) the influence of observations. They suggest that the deviance residuals are more useful in identifying the observations which may be outliers.

Example 3.5.10
Using the semiconductor integrated circuit data as in Example 3.4.1 and

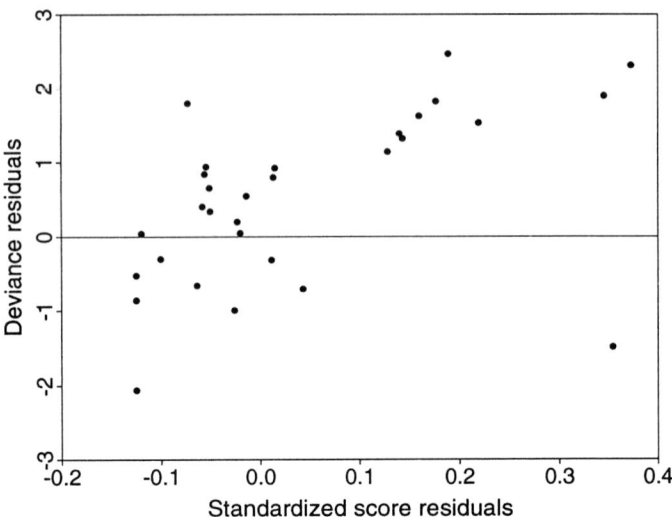

Fig. 3.10. Plot of the deviance residuals versus the standardized score residuals for the inverse of the absolute temperature of the proportional hazards model for the semiconductor integrated circuit data

fitting the proportional hazards model, the martingale residuals are calculated and are presented in Fig. 3.9 as a plot of the residuals against the covariate, the inverse of the absolute temperature. A non-parametric estimate of the regression line, the 'lowess' line, is included. There is some suggestion that a linear fit might not be best and a quadratic function might be an improvement, but this was found not to be the case.

The deviance residuals are calculated and can be used with the standardized score residuals to identify outliers. The plot of these residuals in Fig. 3.10 suggests that there are some outliers with standardized score residuals which are larger than 0.3.

□

3.6　Non-proportional hazards models

An assumption of the Cox regression model was that the hazards are proportional. This can be interpreted as the distance between the $\log(-\log)$ of the reliability functions not varying with time. Cox (1972) suggested a test for this proportionality by adding an extra covariate of log time to the model. In the case of two groups with z_i indicating membership of a group ($z_i = 0$ if the ith individual belongs to group 1, $z_i = 1$ belongs to group 2), then the hazard functions become

$$\text{for group 1:} \lambda_0(t)$$

$$\text{for group 2:} \lambda_0(t) t^{\beta_2} \exp(\beta_1 z_i).$$

If the coefficient of log time is significantly different from zero the reliability functions are non-proportional, otherwise they are proportional. There was concern about the use of such models with time-dependent covariates, though this was justified by Andersen and Gill (1982).

Example 3.6.1
Using the sodium sulphur battery data in Example 3.3.1 and fitting Cox's non-proportional hazards model, $\beta_2 = 0.0900$ with standard error 0.523. These give a Wald statistic of 0.0296, which is not significantly different from zero, for a χ^2 distribution with 1 degree of freedom. Hence the null hypothesis of proportionality of the hazards is accepted, which agrees with the conclusion made in Example 3.5.8, when using the Schoenfeld residuals.

□

Many authors have extended this use of time-dependent variables in studies of lifetimes, for example Gore *et al.* (1984). A number of models have been proposed for the nature of dependency. Some of the models can lead to computational difficulties, see Holford (1976) and Anderson and Senthilselvan (1982). The other obvious problem is selection of an appropriate model. A plot of $\log(-\log)$ of the reliability function or a log hazard plot may reveal some suitable function or the nature of the problem may suggest some particular structure. Alternatively one might use the smoothed martingale residuals suggested by Therneau *et al.* (1990) plotted against the covariate to seek functional form. The danger is in producing too complex a model which does not increase insight. When the variables are quantitative rather than qualitative then the problem is exacerbated and greater caution is necessary.

When predicting future performance there may be difficulties with time-dependent covariates. It may then be necessary to resort to simulation.

3.7 Logistic model

Besides the models described so far there are a number of other models which are either specific to a type of application, such as growth models, see Chapter 7, or specific to the type of data available. In this section only the logistic regression model is considered as an example of these other models. There are many alternatives; see Cox and Oakes (1984) and Lawless (1982) for further examples.

In this model it is assumed that the probability of failure in a specified period is dependent on the covariates $\mathbf{z}$. The model is then

$$P(\text{failure} \mid \mathbf{z}; \boldsymbol{\beta}) = \frac{\exp(\boldsymbol{\beta}^T \mathbf{z})}{1 + \exp(\boldsymbol{\beta}^T \mathbf{z})} .$$

Solutions can be obtained by using maximum likelihood and many statistical packages, such as SAS, GLIM, and S-PLUS, facilitate logistic regression modelling. The likelihood for data in which there are m failures out of n trials is

$$L(\boldsymbol{\beta}) = \frac{\prod_{i=1}^{m} \exp(\boldsymbol{\beta}^T \mathbf{z}_i)}{\prod_{i=1}^{n} (1 + \exp(\boldsymbol{\beta}^T \mathbf{z}_i))} ,$$

where $\mathbf{z}_i$ are the covariates for the ith failure. The first differentials of the log likelihood are

$$\frac{\partial l(\boldsymbol{\beta})}{\partial \beta_j} = \sum_{i=1}^{m} z_{ij} - \sum_{i=1}^{n} \frac{z_{ij} \exp(\boldsymbol{\beta}^T \mathbf{z}_i)}{(1 + \exp(\boldsymbol{\beta}^T \mathbf{z}_i))} ,$$

for $j = 1, 2, \ldots, r$. These derivatives can be equated to zero and solved to give estimators $\hat{\boldsymbol{\beta}}$ of $\boldsymbol{\beta}$.

The second differentials are

$$\frac{\partial^2 l}{\partial \beta_j \partial \beta_k} = -\sum_{i=1}^{n} \frac{z_{ij} z_{ik} \exp(\boldsymbol{\beta}^T \mathbf{z}_i)}{(1 + \exp(\boldsymbol{\beta}^T \mathbf{z}_i))} + \sum_{i=1}^{n} \frac{z_{ij} z_{ik} \exp(2\boldsymbol{\beta}^T \mathbf{z}_i)}{(1 + \exp(\boldsymbol{\beta}^T \mathbf{z}_i))^2} .$$

These second derivatives can be evaluated at $\hat{\boldsymbol{\beta}}$ to produce the observed information matrix whose inverse can be used to obtain standard errors for $\hat{\boldsymbol{\beta}}$.

Example 3.7.1

Consider the sodium sulphur battery data in Example 3.3.1 with the covariate definition for $\mathbf{z}_i$. Suppose the information had been recorded only in terms of whether the battery was working at 1000 cycles. Then the data would be as follows, where 1 denotes a failure up to 1000 cycles and 0 denotes a failure after 1000 cycles:

Batch 1 : 1 1 1 1 1 1 1 1 1 1 0 0 0 0 0 0

Batch 2 : 1 1 1 1 1 1 1 1 1 1 1 1 0 0 0 0 0 0 0 0

Fitting the model

$$P(\text{failure up to } 1000 \mid \text{batch}) = \frac{\exp(\beta_1 + \beta_2 z)}{(1 + \exp(\beta_1 + \beta_2 z))}$$

to the data by maximum likelihood estimation gives $\hat{\beta}_1 = 0.693$ and $\hat{\beta}_2 = -0.154$ with an asymptotic variance covariance matrix

$$\begin{pmatrix} 0.298 & -0.298 \\ -0.298 & 0.523 \end{pmatrix}.$$

Again it is possible to use asymptotic results to test whether the two batches differ, which is equivalent to testing whether $\beta_2 = 0$ or not. The Wald statistic is 0.048, which is not significantly different from zero, for a χ^2 distribution with 1 degree of freedom. Hence the null hypothesis of no difference between the batches is accepted.

□

Examination of the appropriateness of the model can be carried out through the residuals. The logistic regression model is discussed in a number of texts including Lawless (1982). More details of the logistic regression model and other models were given by McCullagh and Nelder (1989).

3.8 Selecting the model and the variables

The number of variables included in an analysis should be considerably less than the number of data points. Overfitting the data can be a major problem. Adding extra variables should improve the fit of the model to the data. However, this will not necessarily improve the ability of the model to predict future observations. Some authors have suggested that data in regression analyses should be split into two parts. With one half one derives the model and with the second half judgements about the model should be made. Often there is too little data to allow such an approach. In the case of limited data one should be wary therefore of too good a fit to the data.

Most of the assessment of which variable to fit will be based on the analysis performed. In normal regression modelling considerable attention has been paid to variable selection, though no generally accepted methodology has been devised to obtain the best set. Two approaches taken are forward selection in which variables are added to the model and backward selection in which variables are deleted until a 'good' model is fitted. These approaches can be applied together allowing for the inclusion and deletion of variables.

For the models considered in this chapter an equivalent to the F-value, used for models with normal errors, would be a change in *deviance*, which for the ith variable is

$$\text{Dev} = -2(l(\beta_i = 0) - l(\beta_i = 0)),$$

where $l(\beta_i = 0)$ is the log likelihood without the ith variable fitted and $l(\beta_i = 0)$ is with it fitted. This statistic is asymptotically distributed as χ^2 with 1 degree of freedom. Hence the addition, or deletion, of a variable given the current model can be decided on whether this value is significantly large or not. Whether the approach results in the 'best' fit is always doubtful, but if there are a large number of variables it may be a sensible approach.

Again if the above method is used then the model produced should be capable of physical explanation within the context. The model ought to seem plausible to others. If the model cannot be interpreted then there is a danger of treating the technique as a black box.

3.9   Discussion

Given the variety of models discussed it may seem crucial that the appropriate model is selected. Whilst in some specific cases this may be the case, generally the desire is to identify factors or variables which are having an effect on the component's or system's performance. In such cases most of the models will at least be able to detect the effects of factors which are strongly associated with lifetimes, though they may overlook those with weak association provided the proportion of censored observations is small. Solomon (1984) has also supported these comments when comparing accelerated failure time and proportional hazards models. Further work is necessary in the case of highly censored data. Hence model selection may not be as crucial as initially suggested. Obviously if the desire is to model lifetimes of the component in detail then more care in choice should be taken.

In this chapter the appropriateness of the models in terms of plots of the residual analysis has been discussed. An alternative judgement on whether a model is appropriate could be based on a 'goodness-of-fit' test. Hence if residuals should come from a given distribution one can apply an appropriate 'goodness-of-fit' test. Such material is outside the scope of this text. Those wanting information on such tests should consult D'Agostino and Stephens (1986).

4
System reliability

4.1 Introduction

This chapter introduces the basic ideas for analysing system reliability. In Chapters 2 and 3 methods of analysing single components or systems were presented and no account was taken of the construction of the systems from components (or subsystems). Often there is information on the performance of these components (or subsystems) and it is sensible to use this information in deciding on the reliability of the system. If data are available on the performance of components it is possible to calculate (estimate) the reliability of these components. Then it is necessary to have methods for calculating (estimating) the reliability of the system constructed from these components. The basic ideas and methods are presented in this chapter.

Firstly, the composition (structure) of the system must be defined in terms of its components. This is done using the structure function with the associated ideas of paths and cuts. These are illustrated using simple examples of series and parallel systems, as well as the important k-out-of-n system. The structure function can be used for probability calculations to obtain the reliability of the system. (The reliability function has been introduced as a function of time, t. In this chapter it will be assumed that a particular value of t is of interest, and this will not be changed during the calculations. Hence, for this reason, the time notation will not be used in this chapter.) For simplicity the probabilistic results are restricted to systems with coherent structure functions, which is all that is needed for most practical applications. If the components operate independently and have identical probabilistic mechanisms then the calculation of the reliability of the system is simplified and a number of results follow. One of the most important is that the reliability function of the system of n components is a linear combination of the reliability function of k-out-of-n systems. As these calculations are by no means as easy for non-identical components there is a real need for good methods which can be used to give bounds on the system reliability. The important topic of three-state reliability, which covers the problem of two kinds of failure, is expounded.

Finally the chapter is concluded with the topics of fault tree analysis, fault modes and effect analysis, and influence diagrams.

4.2 Structure functions

Let a system have n components and let x_i denote the state of the ith component, where

$$x_i = \begin{cases} 1, & \text{if the component is operating,} \\ 0, & \text{if the component has failed.} \end{cases}$$

Then the state of the system is determined from the vector of component states $\mathbf{x} = (x_1, x_2, ..., x_n)$ by the *structure* function $\phi(\mathbf{x})$ which will take the value 1 or 0 respectively when the system is operating or has failed. Note that $x_i^2 = x_i$.

A vector $\mathbf{x}$ for which $\phi(\mathbf{x}) = 1$ is called a *path* and a vector $\mathbf{x}$ for which $\phi(\mathbf{x}) = 0$ is called a *cut* for the structure function $\phi(\mathbf{x})$. As a vector is either a path or a cut the total number of paths and cuts is 2^n.

The number of components which are operating, when the state of the system is determined by $\mathbf{x}$, is called the *size*, $s(\mathbf{x})$, of $\mathbf{x}$, so that

$$s(\mathbf{x}) = \sum_{i=1}^{n} x_i.$$

The number of paths of size k, say, will be denoted by A_k, and hence the number of cuts of size k is $\binom{n}{k} - A_k$.

A path $\mathbf{x}$ is a *minimal path* if $\phi(\mathbf{x}) = 1$ but for every $\mathbf{y} < \mathbf{x}$ then $\phi(\mathbf{y}) = 0$. So if any one of the components which were operating for the path $\mathbf{x}$ were to fail then the system must fail. So minimal paths give the configurations of the minimum number of components required to operate for the system to operate. Hence as long as the components corresponding to at least one of the minimal paths operate then the system operates. This fact can be used to obtain the structure function of the system. There is a corresponding (dual) definition for a *minimal cut*.

For ease of presentation a path will be denoted by the components which equal 1. So, for example, for $n = 5$ the path $(1, 0, 1, 0, 0)$ will be denoted by $x_1 x_3$, and will have a size of 2.

Example 4.2.1
A *series* system is a system in which all components must operate for the system to operate. A two-component series system is illustrated in Fig. 4.1

Fig. 4.1. Two-component series system

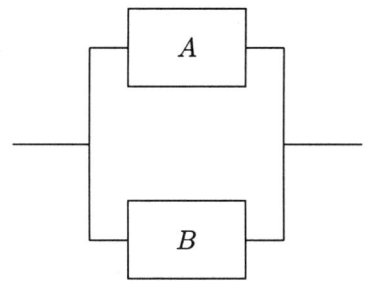

Fig. 4.2. Two-component parallel system

using a *block diagram*. So all the x_i must equal 1, so that $\mathbf{x} = \mathbf{1}$, for $\phi(\mathbf{x})$ to equal 1 and

$$\phi(\mathbf{x}) = \prod_{i=1}^{n} x_i.$$

Hence there is only one path $x_1 x_2 x_3 ... x_n$, which has size n, and hence this is the minimal path.

A *parallel* system is a system in which only one component needs to operate for the system to operate. A two-component parallel system is illustrated in Fig. 4.2 using a *block diagram*. So one of the x_i must equal 1 for $\phi(\mathbf{x})$ to equal 1 and

$$\phi(\mathbf{x}) = 1 - \prod_{i=1}^{n}(1 - x_i).$$

There are $2^n - 1$ paths, as every state of size greater than or equal to 1 will be a path, as there is only one cut, the zero vector $\mathbf{x} = \mathbf{0}$. The minimal paths are all the n paths of size 1.

A *k-out-of-n* system is a system in which at least k components must operate for the system to operate. So at least k of the x_i must equal 1 for $\phi(\mathbf{x})$ to equal 1 and

$$\phi(\mathbf{x}) = \begin{cases} 1, & \text{if } \sum x_i \geq k, \\ 0, & \text{if } \sum x_i < k. \end{cases}$$

Every state of size $j \geq k$ will be a path and there will be $\binom{n}{j}$ paths of size j, so that $A_j - \binom{n}{j}$. Otherwise, if $j < k$ then $A_j = 0$. The minimal paths are all the $\binom{n}{k}$ paths of size k.

It is not possible in general to illustrate a k-out-of-n system using a block diagram, though the series system is an n-out-of-n system and the parallel system is a 1-out-of-n system.

□

Before considering any further examples it will be useful to give some results based on the examples of series and parallel systems. It is seen, in general, that if two systems with structure functions $\phi_1(\mathbf{x}_1)$ and $\phi_2(\mathbf{x}_2)$, respectively, are put in series then the structure function of the resulting system is given by

$$\phi(\mathbf{x}) = \phi_1(\mathbf{x}_1)\phi_2(\mathbf{x}_2),$$

and if put in parallel then the structure function of the resulting system is given by

$$\phi(\mathbf{x}) = 1 - (1 - \phi_1(\mathbf{x}_1))(1 - \phi_2(\mathbf{x}_2))$$

where $\mathbf{x} = (\mathbf{x}_1, \mathbf{x}_2)$. In general the components of $\mathbf{x}_1$ and $\mathbf{x}_2$ will be distinct as the two systems will consist of different components. However, in representing more complicated systems it will be convenient to combine systems which are made up by components from the minimal paths or cuts and these systems may have components in common. It will be useful to know how to simplify the resulting structure function and this can be done using the following two rules.

Rule 4.2.1

If the structure functions of the two systems are $\phi_1(\mathbf{x}) = \phi_0(\mathbf{x})\phi_A(\mathbf{x})$ and $\phi_2(\mathbf{x}) = \phi_0(\mathbf{x})\phi_B(\mathbf{x})$, respectively, and are put in series then the structure function of the resulting system is given by

$$\begin{aligned} \phi(\mathbf{x}) &= \phi_1(\mathbf{x})\phi_2(\mathbf{x}) \\ &= \phi_0(\mathbf{x})\phi_A(\mathbf{x})\phi_B(\mathbf{x}), \end{aligned}$$

as $\phi_0(\mathbf{x})^2 = \phi_0(\mathbf{x})$.

Rule 4.2.2

If the structure functions of the two systems are $\phi_1(\mathbf{x}) = \phi_0(\mathbf{x})\phi_A(\mathbf{x})$ and $\phi_2(\mathbf{x}) = \phi_0(\mathbf{x})\phi_B(\mathbf{x})$, respectively, and are put in parallel then the structure function of the resulting system is given by

$$\phi(\mathbf{x}) = 1 - (1 - \phi_1(\mathbf{x}))(1 - \phi_2(\mathbf{x}))$$

$$= \phi_0(\mathbf{x})(1 - (1 - \phi_A(\mathbf{x}))(1 - \phi_B(\mathbf{x}))),$$

as $\phi_0(\mathbf{x})^2 = \phi_0(\mathbf{x})$.

Example 4.2.2
Consider the example in Fig. 4.3 of five components A, B, C, D, and E. Let the vector $\mathbf{x}$ represent the state of the system with the order of the components corresponding to the order of the letters of the components and let the structure function be $\phi(\mathbf{x})$.

The paths of size 2 are $x_1 x_4$ and $x_1 x_5$;
of size 3 are $x_1 x_2 x_4$, $x_1 x_2 x_5$, $x_1 x_3 x_4$, $x_1 x_3 x_5$, $x_1 x_4 x_5$, $x_2 x_3 x_4$, and $x_2 x_3 x_5$;
of size 4 $x_1 x_2 x_3 x_4$, $x_1 x_2 x_3 x_5$, $x_1 x_2 x_4 x_5$, $x_1 x_3 x_4 x_5$, and $x_2 x_3 x_4 x_5$; and
of size 5 $x_1 x_2 x_3 x_4 x_5$.

Hence the path numbers for this system are $A_1 = 0$, $A_2 = 2$, $A_3 = 7$, $A_4 = 5$, and $A_5 = 1$.

The minimal paths are $x_1 x_4$, $x_1 x_5$, $x_2 x_3 x_4$, and $x_2 x_3 x_5$.

The system will operate if the components corresponding to at least one of these minimal paths operate. Hence applying the method used in finding the structure function of a parallel system gives the structure function as

$$\begin{aligned} \phi(\mathbf{x}) &= 1 - (1 - x_1 x_4)(1 - x_1 x_5)(1 - x_2 x_3 x_4)(1 - x_2 x_3 x_5) \\ &= x_1 x_4 + x_1 x_5 - x_1 x_4 x_5 + x_2 x_3 x_4 + x_2 x_3 x_5 \\ &\quad - x_1 x_2 x_3 x_4 - x_1 x_2 x_3 x_5 - x_2 x_3 x_4 x_5 + x_1 x_2 x_3 x_4 x_5, \end{aligned}$$

using the result that $x_i^2 = x_i$.

However, it is possible to rearrange this structure function as a product to give

$$\phi(\mathbf{x}) = (x_1 + x_2 x_3 - x_1 x_2 x_3)(x_4 + x_5 - x_4 x_5).$$

This is done by using Rule 4.2.2 on

$$(1 - x_1 x_4)(1 - x_2 x_3 x_4) = 1 - (1 - (1 - x_1 x_4)(1 - x_2 x_3 x_4))$$

with $\phi_0(\mathbf{x}) = x_4$, and on $(1 - x_1 x_5)(1 - x_2 x_3 x_5)$ with $\phi_0(\mathbf{x}) = x_5$, with in both cases $\phi_A(\mathbf{x}) = x_1$ and $\phi_B(\mathbf{x}) = x_2 x_3$. Then Rule 4.2.2 is applied again on the resulting structure function using $\phi_0(\mathbf{x}) = x_1 + x_2 x_3 - x_1 x_2 x_3$ with $\phi_A(\mathbf{x}) = x_4$ and $\phi_B(\mathbf{x}) = x_5$, which gives the structure function in the form of a product of a function of x_1, x_2, and x_3 and a function of x_4 and x_5.

Alternatively, rather than use the minimal paths, it is possible to use the minimal cuts to obtain the structure function of the system. The minimal cuts are $x_1 x_2$, $x_1 x_3$, and $x_4 x_5$.

The system will not operate if the components corresponding to at least one of the minimal cuts do not operate. Hence applying the method used in finding the structure function of a series system gives the structure function as

$$
\begin{aligned}
\phi(\mathbf{x}) &= (x_1 + x_2 - x_1 x_2)(x_1 + x_3 - x_1 x_3)(x_4 + x_5 - x_4 x_5) \\
&= (x_1 + x_2 x_3 - x_1 x_2 x_3)(x_4 + x_5 - x_4 x_5),
\end{aligned}
$$

by using Rule 4.2.2 on

$$
(x_1 + x_2 - x_1 x_2)(x_1 + x_3 - x_1 x_3) = (1 - (1 - x_1)(1 - x_2))(1 - (1 - x_1)(1 - x_3))
$$

with $\phi_0(\mathbf{x}) = 1 - x_1$, $\phi_A(\mathbf{x}) = 1 - x_2$, and $\phi_B(\mathbf{x}) = 1 - x_3$.

From these results it is seen that for simple systems it may be easier to obtain the structure function by building up the system using series and parallel structures. The structure function for components B and C, which are in series, is $x_2 x_3$ and hence the structure function for components A, B, and C, as A is in parallel with B and C, is $x_1 + x_2 x_3 - x_1 x_2 x_3$. The structure function for components D and E, which are in parallel, is $x_4 + x_5 - x_4 x_5$. The system is then formed by putting the system formed by components A, B, and C in series with the system formed by components D and E, which by taking the product of the two structure functions gives the structure function obtained above.

□

From this example it is seen that using the minimal paths and cuts will lead initially to structure functions which may differ *algebraically*, though of course they are *numerically* equivalent. This is because x_i raised to any power is *numerically* equal to x_i.

4.3 Probability calculations for coherent systems

It is often the case that only systems with *coherent* structure functions are considered and so the definition for such structures will now be given. There may be cases where *non-coherent* structure functions arise but these are the exception. To do this two new definitions are required. Firstly the ith component is *irrelevant* if

$$
\phi(x_1, ..., x_{i-1}, 0, x_{i+1}, ..., x_n) = \phi(x_1, ..., x_{i-1}, 1, x_{i+1}, ..., x_n),
$$

so that the state of the system is the same whether x_i is 1 or 0. A component which is not irrelevant is *relevant*. Secondly the structure function is *monotonic* if for every i

$$\phi(x_1, ..., x_{i-1}, 0, x_{i+1}, ..., x_n) \leq \phi(x_1, ..., x_{i-1}, 1, x_{i+1}, ..., x_n).$$

Definition of a coherent structure function
A system has a coherent structure function if the function is monotonic and every component is relevant. Birnbaum *et al.* (1961) gave an account of these results.

If the n components of a system with a coherent structure function are *independent* and p_i is the reliability of the ith component (the probability that the ith component is operating) then the reliability of the system is given by

$$
\begin{aligned}
R(\mathbf{p}) &= E(\phi(\mathbf{x})) \\
&= \sum_{\mathbf{x}} [\phi(\mathbf{x}) \prod_{i=1}^{n} \{p_i^{x_i}(1 - p_i)^{1-x_i}\}].
\end{aligned}
$$

Example 4.3.1
For a *series* system

$$R(\mathbf{p}) = \prod_{i=1}^{n} p_i,$$

as there is only one path, and for a *parallel* system

$$R(\mathbf{p}) = 1 - \prod_{i=1}^{n}(1 - p_i),$$

using the fact that there is only one cut.

□

Before considering any further examples it will be useful to give some results based on the examples of series and parallel systems. It is seen, in general, that if two systems with structure functions $\phi_1(\mathbf{x}_1)$ and $\phi_2(\mathbf{x}_2)$, respectively, where $\mathbf{x} = (\mathbf{x}_1, \mathbf{x}_2)$, are put in series then the structure function of the resulting system is given by

$$\phi(\mathbf{x}) = \phi_1(\mathbf{x}_1)\phi_2(\mathbf{x}_2),$$

and if put in parallel then the structure function of the resulting system is given by
$$\phi(\mathbf{x}) = 1 - (1 - \phi_1(\mathbf{x}_1))(1 - \phi_2(\mathbf{x}_2)).$$

Let the components of $\mathbf{x}_1$ and $\mathbf{x}_2$ be distinct so that the two systems consist of different components and are independent.

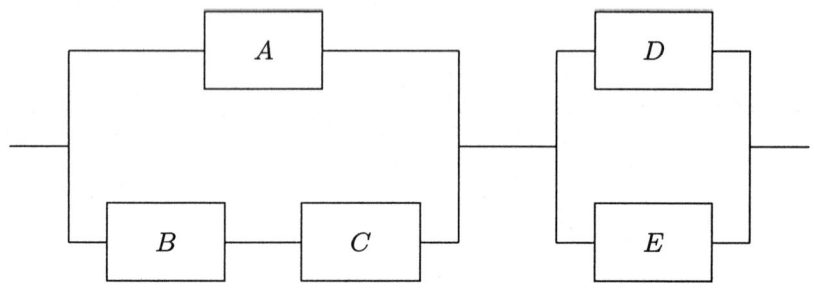

Fig. 4.3. Five-component system

Rule 4.3.1
If the two systems are put in series then the reliability of the resulting
system is given by

$$\begin{aligned} R(\mathbf{x}) &= E(\phi_1(\mathbf{x}_1)\phi_2(\mathbf{x}_2)) \\ &= R_1(\mathbf{x})R_2(\mathbf{x}). \end{aligned}$$

Rule 4.3.2
If the two systems are put in parallel then the reliability of the resulting
system is given by

$$\begin{aligned} R(\mathbf{x}) &= E(1 - (1 - \phi_1(\mathbf{x}))(1 - \phi_2(\mathbf{x}))) \\ &= 1 - (1 - R_1(\mathbf{x}))(1 - R_2(\mathbf{x})) \\ &= R_1(\mathbf{x}) + R_2(\mathbf{x}) - R_1(\mathbf{x})R_2(\mathbf{x}). \end{aligned}$$

Example 4.3.2
Consider the example in Fig. 4.3 of five components A, B, C, D, and E.
The structure function is given by

$$\phi(\mathbf{x}) = (x_1 + x_2 x_3 - x_1 x_2 x_3)(x_4 + x_5 - x_4 x_5).$$

Apply Rule 4.3.1 by putting B and C in series to give a reliability of
$p_2 p_3$, and then apply Rule 4.3.2 by putting the resulting system in parallel

with A to give a reliability function of $(p_1 + p_2p_3 - p_1p_2p_3)$. Apply Rule 4.3.2 by putting D in parallel with E which will give a reliability function of $(p_4 + p_5 - p_4p_5)$. Finally apply Rule 4.3.1 by putting these two systems in series. Hence the reliability of the total system is given by

$$
\begin{aligned}
R(\mathbf{p}) &= \phi(\mathbf{p}) \\
&= (p_1 + p_2p_3 - p_1p_2p_3)(p_4 + p_5 - p_4p_5).
\end{aligned}
$$

$\square$

So in these examples it is seen that $R(\mathbf{p}) = \phi(\mathbf{p})$. This is because $\phi(\mathbf{x})$ is *linear* in all the variables. In general this method does not enable the reliability to be calculated easily if this is not the case.

4.4   Reliability for systems of identical components

Let the n components of a system with a coherent structure function be *independent* and the reliability of the ith component (the probability that the ith component is operating) be equal to p, so that all the components are *identical*. Then the reliability of the system is given by

$$
\begin{aligned}
R(\mathbf{p}) &= E(\phi(\mathbf{x})) \\
&= \sum_{i=1}^{n} A_i p^i (1 - p)^{n-i}.
\end{aligned}
$$

This simplifies the reliability of the system as it is a one-dimensional function depending only on p.

Example 4.4.1
For a *series* system
$$
R(\mathbf{p}) = p^n,
$$
as there is only one path, and for a *parallel* system
$$
R(\mathbf{p}) = 1 - (1 - p)^n,
$$
using the fact that there is only one cut. For a k-out-of-n system $A_j = \binom{n}{j}$, for $j \geq k$, and $A_j = 0$, otherwise, and hence the reliability is given by

$$
R(\mathbf{p}) = \sum_{i=k}^{n} \binom{n}{i} p^i (1 - p)^{n-i}.
$$

$\square$

The last result for the k-out-of-n system shows that the reliability of the system of n components can be given in terms of the reliability of k-out-of-n systems. This is because

$$R(\mathbf{p}) = \sum_{i=1}^{n} A_i p^i (1-p)^{n-i}$$

$$= \sum_{i=1}^{n} \left[\sum_{j=1}^{i} w_j \right] \binom{n}{i} p^i (1-p)^{n-i},$$

$$\text{where } \sum_{j=1}^{i} w_j = A_i / \binom{n}{i},$$

$$= \sum_{j=1}^{n} w_j \left[\sum_{i=j}^{n} \binom{n}{i} p^i (1-p)^{n-i} \right]$$

$$= \sum_{j=1}^{n} w_j R(\mathbf{p})_{j|n},$$

where $R(\mathbf{p})_{j|n}$ is the reliability of the j-out-of-n system. Note that $w_j \geq 0$ and $\sum_{j=1}^{n} w_j = 1$, so that the reliability of the system is a *linear combination* of the n reliability of the k-out-of-n systems. This result will be useful later when two kinds of failure are considered.

Example 4.4.2
Consider the example in Fig. 4.3 of five components A, B, C, D, and E. The path numbers for this system are $A_1 = 0$, $A_2 = 2$, $A_3 = 7$, $A_4 = 5$, and $A_5 = 1$. Using the path numbers, the reliability is given by

$$R(\mathbf{p}) = \sum_{i=1}^{n} A_i p^i (1-p)^{n-i}$$

$$= 2p^2(1-p)^3 + 7p^3(1-p)^2 + 5p^4(1-p) + p^5.$$

The structure function is given by

$$\phi(\mathbf{x}) = (x_1 + x_2 x_3 - x_1 x_2 x_3)(x_4 + x_5 - x_4 x_5).$$

Hence the reliability of the system is given by

$$R(\mathbf{p}) = \phi(\mathbf{p})$$

$$= (p + p^2 - p^3)(2p - p^2).$$

Both these formulae reduce to

$$R(\mathbf{p}) = 2p^2 + p^3 - 3p^4 + p^5.$$

From the path numbers $w_1 = 0$, $w_2 = 1/5$, $w_3 = 1/2$, $w_4 = 3/10$, and $w_5 = 0$. Hence

$$R(\mathbf{p}) = \frac{1}{5}[R(\mathbf{p})_{2|5}] + \frac{1}{2}[R(\mathbf{p})_{3|5}] + \frac{3}{10}[R(\mathbf{p})_{4|5}].$$

$\square$

4.5 Bounds on system reliability

The n components of a system with a coherent structure function are *independent* and p_i is the reliability of the ith component (the probability that the ith component is operating). Let the set of minimal paths of the system be denoted by $\mathcal{A}$ and the set of minimal cuts of the system be denoted by $\mathcal{C}$. Then it can be shown that the following inequality holds for the reliability of the system:

$$\prod_{\mathbf{x} \in \mathcal{C}} \left\{ 1 - \prod_{i=1}^{n} (1 - p_i)^{1 - x_i} \right\} \le R(\mathbf{p}) \le 1 - \left\{ \prod_{\mathbf{x} \in \mathcal{A}} \left(1 - \prod_{i=1}^{n} p_i^{x_i} \right) \right\}.$$

Example 4.5.1
Consider the example in Fig. 4.3 of five components A, B, C, D, and E.
The minimal paths are $x_1 x_4$, $x_1 x_5$, $x_2 x_3 x_4$, and $x_2 x_3 x_5$.
The minimal cuts are $x_1 x_2$, $x_1 x_3$, and $x_4 x_5$.
Hence the reliability of the system satisfies the inequalities given by

$$(p_1 + p_2 - p_1 p_2)(p_1 + p_3 - p_1 p_3)(p_4 + p_5 - p_4 p_5)$$
$$\le R(\mathbf{p}) = (p_1 + p_2 p_3 - p_1 p_2 p_3)(p_4 + p_5 - p_4 p_5)$$
$$\le 1 - (1 - p_1 p_4)(1 - p_2 p_3 p_4)(1 - p_1 p_5)(1 - p_2 p_3 p_5).$$

If all the items are identical, so that $p_i = p$, then the inequalities become

$$(2p - p^2)^3 \le$$
$$R(\mathbf{p}) = (p + p^2 - p^3)(2p - p^2)$$
$$\le 1 - (1 - p^2)^2 (1 - p^3)^2.$$

A plot of the system reliability and the upper and lower bounds is given in Fig. 4.4.

$\square$

It is seen from Fig. 4.4 that the system reliability is 'S-shaped', which is the case for most practical systems, see Birnbaum *et al.* (1961).

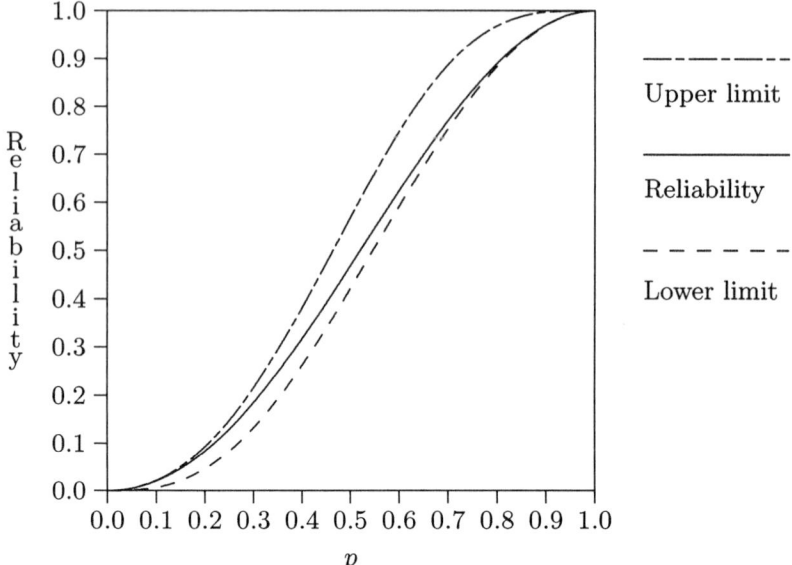

Fig. 4.4. The reliability of the five-component system given in Fig. 4.3 for identical components with upper and lower bounds

4.6   Approximations

These results given in Section 4.5 suggest some methods which may be used for approximating the reliability of a system. These methods may be useful when dealing with complicated systems. It was seen in Fig. 4.4 when considering the system used in Example 4.5.1 that, at both ends of the interval $[0,1]$ for p, the component reliability, one or other of the bounds on system reliability was very good. For p near 0 the upper bound was good while for p near 1 the lower bound was good. This immediately shows that if an approximation is to be used then it is likely to be good for p either near 0, for very 'unreliable' components, or near 1, for very 'reliable' components.

The first method is to use the lowest degree terms in p in the equation of the system reliability to give an approximation near 0. Alternatively if an approximation is required near 1 then express the system reliability as a function of $1 - p$ and use 1 and the lowest degree terms in $1 - p$.

The second method is to use the bounds on system reliability given in Section 4.5. As was seen in Fig. 4.4 it is best to use the upper bound if p is near 0 and the lower bound if p is near 1.

The third method is to use a combination of the two bounds on system reliability. As it is best to use the upper bound if p is near 0 and the lower bound if p is near 1, this suggests that a weighting of these two bounds using a weight of p on the lower limit and a weight of $1 - p$ on the upper limit would be best. This approximation will behave like the upper limit for p near 0 and like the lower limit for p near 1 as is desired.

Example 4.6.1

Consider the example in Fig. 4.3 of five components A, B, C, D, and E. The reliability of the system is given by

$$R(\mathbf{p}) = (p_1 + p_2 p_3 - p_1 p_2 p_3)(p_4 + p_5 - p_4 p_5).$$

The first method would use the approximation

$$R(\mathbf{p}) \approx p_1(p_4 + p_5),$$

for $\mathbf{p}$ near $\mathbf{0}$, which reduces to

$$R(p) \approx 2p^2$$

for identical components.

For $\mathbf{p}$ near $\mathbf{1}$ the first method would use the approximation

$$R(\mathbf{p}) \approx 1 - (1 - p_1)(1 - p_2) - (1 - p_1)(1 - p_3) - (1 - p_4)(1 - p_5),$$

which reduces to

$$R(p) \approx 1 - 3(1 - p)^2 = -2 + 6p - 3p^2$$

for identical components.

The second method would use the approximation

$$R(\mathbf{p}) \approx 1 - (1 - p_1 p_4)(1 - p_2 p_3 p_4)(1 - p_1 p_5)(1 - p_2 p_3 p_5),$$

for $\mathbf{p}$ near $\mathbf{0}$, which reduces to

$$R(p) \approx 2p^2 + 2p^3 - p^4 - 4p^5 - p^6 + 2p^7 + 2p^8 - p^{10}$$

for identical components.

For $\mathbf{p}$ near $\mathbf{1}$ the second method would use the approximation

$$R(\mathbf{p}) \approx (p_1 + p_2 - p_1 p_2)(p_1 + p_3 - p_1 p_3)(p_4 + p_5 - p_4 p_5),$$

which reduces to

$$R(p) \approx 1 - 3(1 - p)^2 + 3(1 - p)^4 - (1 - p)^6 = 8p^3 - 12p^4 + 6p^5 - p^6$$

for identical components.

Table 4.1. The reliability of the five-component system given in Fig. 4.3 for identical components with approximations

Component reliability p	System reliability $R(p)$	First approximation	Second approximation	Third approximation
0.00	0.00000	0.00000	0.00000	0.00000
0.05	0.00511	0.00500	0.00524	0.00503
0.10	0.02071	0.02000	0.02186	0.02036
0.15	0.04693	0.04500	0.05093	0.04650
0.20	0.08352	0.08000	0.09309	0.08380
0.25	0.12988	0.12500	0.14834	0.13219
0.30	0.18513	0.18000	0.21601	0.19100
0.35	0.24811	0.24500	0.29461	0.25890
0.40	0.31744	0.32000	0.38183	0.33395
0.45	0.39156	0.40500	0.47462	0.41375
0.50	0.46875	0.50000	0.56934	0.49561
0.55	0.54718	0.39250	0.50721	0.57683
0.60	0.62496	0.52000	0.59270	0.65492
0.65	0.70014	0.63250	0.67568	0.72777
0.70	0.77077	0.73000	0.75357	0.79382
0.75	0.83496	0.81250	0.82397	0.85199
0.80	0.89088	0.88000	0.88474	0.90162
0.85	0.93681	0.93250	0.93401	0.94219
0.90	0.97119	0.97000	0.97030	0.97300
0.95	0.99264	0.99250	0.99252	0.99288
1.00	1.00000	1.00000	1.00000	1.00000

The third method would use the approximation

$$R(p) \approx 2p^2 + 5p^4 - 15p^5 + 9p^6 + 2p^7 - 2p^9 - p^{10} + p^{11}$$

for identical components.

The system reliability and the three approximations are given in Table 4.1 for the case of identical components. The approximations obtained by the first and second methods are only used for the appropriate half of the interval, either in [0, 0.5] or (0.5, 1]. Hence there is a discontinuity at $p = 0.5$ and obviously neither of the first two approximations would be suitable for values near $p = 0.5$. The 'poorest' approximation is that given by the first method for values of p in the interval (0.5, 0.8). The second method of approximation is also 'poor' for values of p in the interval (0.2, 0.5). Otherwise the approximations are fairly good. The third method of weighting the two bounds gives the least maximum absolute difference (0.03) between the reliability and the approximation over the relevant part of the interval. However, for this system the first method is the best approximation near 0 and the second method of using the lower bound is the best approximation near 1.

$\square$

It is possible to modify iterative shortest path techniques for calculating system reliability to produce bounds on reliability. The method developed by Shier and Liu (1992) provides not only a numerical answer but also an actual polynomial involving the component reliabilities. Such information can be useful for sensitivity studies to evaluate the effect of changes to the reliabilities of certain components.

4.7   Two kinds of failure

When investigating the reliability of a system it is often necessary to consider not only *failure to operate* but also the fact that the system may operate inadvertently (when it should not do so) and this will be called *failure to idle*. This situation was first considered by Moore and Shannon (1956) and von Neumann (1956). This approach has been applied to systems composed of electric relays or switches, fluid flow valves, and sensors for monitoring or protecting systems.

To ensure the reliability of a system it is often necessary to combine a number of identical and independent components to give a system which contains some redundant components in order to maintain the system in the event of components failing. The system consists of n components which can be in one of two states: they can operate or idle. The components may be subject to two mutually exclusive kinds of failure: the component may fail to operate because the component is fixed permanently in the idling state and the probability of this failure is a; or the component may fail to idle because the component is fixed permanently in the operating state and the probability of this failure is b. Hence the reliability of a component is given by $c = 1 - a - b$. The components are neither totally reliable nor unreliable, i.e. $0 < c < 1$. Hence the system made from n of these components can fail to operate because the system is fixed permanently in the idling state and the probability of this failure is $u(a)$; or the system can

fail to idle because the system is fixed permanently in the operating state
and the probability of this failure is $v(b)$. The system (and any component)
can therefore be considered to be in one of three 'states': failing to operate;
failing to idle; or operating correctly. For this reason many authors refer to
the systems as 'three-state device reliability systems', see Dhillon (1977).

The question of how to determine the state of the system arises. There
are essentially two situations. Firstly it may be possible to obtain complete
information about the system at time t and to know which of three states
the system is in. Secondly it may only be possible to send one 'command'
to the system at time t and then be able to observe whether the system
responds correctly to the 'command'. For example, the 'command' could
be for the components (and the system) to operate and then it is possible
to observe if the system is in either one of the two states of operating
correctly or failing to operate. Alternatively the 'command' could be for
the components (and the system) to idle and then it is possible to observe
if the system is in either one of the two states of operating correctly or
failing to idle.

To obtain the reliability of a system, the probability that the system is
operating correctly, it is necessary to obtain $u(a)$ and $v(b)$. It is usual to
assume that the structure function is coherent, see Section 4.3. Then if the
system is subject to two mutually exclusive kinds of failure the reliability
of the system is given by

$$
\begin{aligned}
R &= 1 - u(a) - v(b) \\
  &= v(1 - a) - v(b),
\end{aligned}
$$

as

$$
u(a) + v(1 - a) = 1.
$$

This is because if all the components are unreliable then so is the system.
Hence in terms of the reliability of a component the reliability of a system
R is given by

$$
R = v(q(1 - c) + c) - v(q(1 - c)),
$$

where $q = b/(a + b)$ is the probability that a failure is a failure to idle.

Example 4.7.1

Consider the two-component series system which is illustrated in Fig. 4.1
for two identical components. From the result in Example 4.4.1 for the
reliability of a series system it is possible to obtain the probability of
the system of independent and identical components failing to operate as
$1 - u(a) = (1 - a)^2$ so that $u(a) = 2a - a^2$.

As the paths for failure to operate for this system are the cuts for failure
to idle for this system then the probability of the system failing to idle can

be obtained using the *dual* system of two components in parallel. Hence $v(b) = b^2$, so that $u(a) + v(1 - a) = 1$. (Hence $v(b) = R(\mathbf{b})$, as given in Example 4.4.1.)

Therefore from the result above the reliability of the system is given by

$$
\begin{aligned}
R &= (q(1 - c) + c)^2 - (q(1 - c))^2 \\
&= 2qc(1 - c) + c^2,
\end{aligned}
$$

where $q = b/(a + b)$ is the probability that a failure is a failure to idle.

Secondly consider the two-component parallel system which is illustrated in Fig. 4.2 for two independent and identical components. As this is the dual of the two-component series system then $v(b) = 1 - (1 - b)^2 = 2b - b^2$, the same function that was used for $u(a)$ for the two-component series system. Hence the reliability of the system is given by

$$
\begin{aligned}
R &= 2(q(1 - c) + c) - (q(1 - c) + c)^2 - 2q(1 - c) + (q(1 - c))^2 \\
&= 2c - c^2 - 2qc(1 - c).
\end{aligned}
$$

$\square$

These simple systems considered in Example 4.7.1 have the property that the two kinds of failure are mutually exclusive for the system state. For these two-component systems, whose components can each be in one of three states, there are nine combinations of component states which the system can take. For all of these nine combinations of states the system will only be in *one* of the three states of failing to operate, failing to idle, or operating correctly.

However, for more complicated systems the two kinds of failure are not necessarily mutually exclusive. This is illustrated by the systems introduced by Jenney and Sherwin (1986), who considered systems with *j or more of m parallel components required*. Such a system fails if more than $m - j$ of the m parallel components fail to operate or if any item fails to idle. This system can both fail to operate and fail to idle simultaneously. Then the above formula for the reliability R is too pessimistic, and gives too small a value for the reliability, as was pointed out by Malon (1989).

Example 4.7.2

Consider the following system discussed by Malon (1989), which is a special case of systems with *j or more of m parallel components required* suggested by Jenney and Sherwin (1986). The system consists of two water pipes which are fixed in parallel, each fitted with a valve to allow through or cut off the flow of water. The valve either operates by allowing the flow of water through the pipe or idles by cutting off the flow of water through the pipe. The system is designed to operate correctly if it is delivering water through **both** pipes, when water is required, and is delivering no water through either pipe, when no water is required. This is an example of a systems with *2 (or more) of 2 parallel components required*.

If the two kinds of failure for this system are failure to operate, the state which is failure to deliver the required water through both pipes, and failure to idle, the state which is failure to cut off the flow of water in both pipes, then it is possible for the system to fail in both states simultaneously. This occurs when one valve fails to operate (fails to allow water to flow through the pipe) and when the other valve fails to idle (fails to stop the flow of water through the pipe).

Hence if the results for the reliability are calculated using the results above for independent and identical components, which assume that the two kinds of failure cannot occur together, the reliability would be given by $1 - 2a + a^2 - 2b + b^2$. This is obtained by using the results obtained in Example 4.7.1 for $u(a)$ for a two-component series system and $v(b)$ for a two-component parallel system. This is less than the correct reliability which is $1 - 2a + a^2 - 2b + b^2 + 2ab$. The difference is $2ab$, the probability that one valve fails to operate and the other valve fails to idle, the situation which leads to the system failing both to operate and to idle.

This illustrates the fact that the results given above and used in Example 4.7.1 must only be used for systems where it is not possible for the system to fail in both states simultaneously. If systems are considered, as was done by Jenney and Sherwin (1986), where it is possible for the system to fail in both states simultaneously, then it is necessary to take into account the probability of this event in the calculation of the reliability of the system. This explains the reason why the results given in Jenney and Sherwin (1986) for systems with this property are incorrect, as they give too small a value for the reliability.

□

Jenney and Sherwin (1986) also considered series systems and systems made up either of parallel groups of components in series or of series groups of components in parallel. These were also examples of systems which can both fail to operate and fail to idle simultaneously. The reliabilities of such systems have been considered by Satoh *et al.* (1993).

For a given number of n components there are a number of possible configurations which can be constructed. Lomnicki (1972) gave the number of different two-terminal parallel–series networks which can be constructed from n identical components. These networks are systems which can be constructed by putting components together in series or parallel. This number increases rapidly with n. Hence it is necessary to be able to choose between a number of systems which belong to a group with a given number of components. In one of the earliest papers to address this problem Moore and Shannon (1956) divided the systems into two groups: a group of 'reliable' systems which had a reliability greater than $1 - 2\epsilon$ as $v(1-a) > 1 - \epsilon$ and $v(b) < \epsilon$; and a group which did not satisfy these conditions. However, the following definition of an optimum system using the reliability of the system was subsequently used by many authors, including Barlow et al. (1963), Barlow and Proschan (1965, 1975), and Phillips (1976).

Definition 4.7.1
For a fixed value of c, a member of a group of systems is defined to be *optimum* if as q varies there exists an interval (q_1, q_2), where $0 \le q_1 < q_2 \le 1$, for which R, the reliability of the system, is larger than the reliabilities of all other systems in the group.

The question arises of how to identify those systems which are optimum among the group of two-terminal parallel–series networks of n components, and their corresponding optimum intervals. According to Jenney and Sherwin (1986) this problem was first solved for $n = 3$ by B.W. Jenney and fellow workers in the late 1950s. However, results do not appear to have been published until the work of Lomnicki (1973) and Phillips (1976). Phillips (1976) gave results for $n = 3, 4$, and 5 and also gave the following theorem which enables systems which are *not* optimum to be eliminated from the group.

Theorem 4.7.1
If a system, which is a member of a group of $l + 1$ systems, has

$$v(b) = \sum_{i=1}^{l} w_i v_i(b),$$

where $v_i(b)$ is the probability of system i in the group failing to idle and the w_i are non-negative weights such that $w_i \ge 0$, $w_i > 0$ for at least two values of i and $\sum_{i=1}^{l} w_i = 1$, then the system is *not* optimum.

$v(b)$ is a *convex combination* of the probabilities of failure to idle of the other systems and hence so is the reliability of the system. Hence for fixed c the system reliability is less than at least one of the other members of the group, and so the system is not optimum. Using Phillips' theorem from Phillips (1976) the problem of finding the optimum systems for $n = 6, 7$, and 8 was solved. Many of these results have subsequently appeared in later papers such as Jenney and Sherwin (1986) and Page and Perry (1988). As a simpler alternative to considering all parallel–series networks it is possible to restrict interest to those systems which are composed of a number of components in series (parallel) put in parallel (series). These systems were considered initially by Price (1960), Barlow and Hunter (1960a), and von Alven (1964) and later by Nakagawa and Hattori (1980).

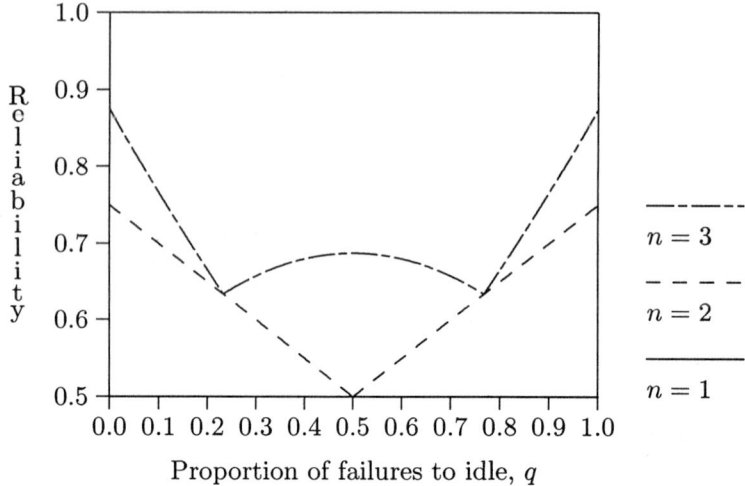

Fig. 4.5. The reliabilities of the optimum systems for the proportion of failures to idle

From the results for identical components in Section 4.4 it was seen that the reliability of a coherent system of n identical and independent components is a convex combination of the reliabilities of the k-out-of-n systems, for $1 \le k \le n$. Hence it was shown by Phillips (1980) that the optimum k-out-of-n systems must be optimum also for the group of coherent systems of n identical and independent components. This result is also true if the components are not independent, as was shown by Ansell and Bendell (1982), provided the components are 'exchangeable', i.e. the probability of k-out-of-n components failing is the same whichever of the k components failed. Hence k-out-of-n systems are 'preferable'.

However, there is still the question of which k-out-of-n system is optimum for any value of q, for a fixed value of c. The k-out-of-n system is optimum for q in an interval $(x_{k-1,n}, x_{k,n})$, where $x_{k-1,n} < x_{k,n}$. Phillips (1980) showed that $x_{0,n} = 0$ and $x_{n,n} = 1$ and $x_{k-1,n} < x_{k-1,n-1} < x_{k,n}$, for $2 \le k \le n - 1$. Hence if

$$U_n(q, c) = \max_{1 \le k \le n} \{R_{k,n}(q, c)\}$$

then

$$U_{n-1}(q, c) \le U_n(q, c), \quad \text{for } n \le 2$$

and
$$U_{n-1}(q, c) < U_n(q, c), \text{ for } q \neq x_{k,n}, \text{ and } 1 \leq k \leq n - 1.$$

The reliability $U_n(q, c)$ of the optimum systems of the group of k-out-of-n systems is given in Fig. 4.5, for $c = \frac{1}{2}$. The question of obtaining the interval of q in which a system is optimum was also considered by Ben-Dov (1980).

Other criteria such as expected time to failure (considered by Barlow and Hunter (1960a), von Alven (1964), and Lomnicki (1973)) or a linear weighting of $u(a)$ and $v(b)$ (considered by Lloyd and Lipow (1962)) instead of reliability will lead to similar conclusions. Sometimes results were obtained which appeared to contradict earlier work. For instance, Kopocinski (1974) considered the case of the limit as $c \to 1$ and concluded that the optimum group of systems obtained was changed from those obtained by Lomnicki (1973). However, all the result showed was that the intervals over which some of the systems were optimum approached zero as c approaches 1. (The situation at $c = 1$ is not of any interest, however, as this is the case of totally reliable components.)

A good survey of the literature on this topic of two kinds of failures is given by Lesanovsky (1993).

4.8   Fault tree analysis and related approaches

Deriving the structure function for a system is usually a two-stage process. The first stage is the analysis of the likely faults and their consequences and the second stage is the production of the mathematical model of the system. In FTA (Fault Tree Analysis) pioneered by Fussel (1976), the first stage is the construction of the fault tree or set of fault trees and the second stage is the analysis of the tree.

Construction of the tree requires engineering knowledge and judgement. The approach taken in fault tree constructions is the top-down procedure. There are alternatives such as FMEA (Fault Modes and Effect Analysis) which is a bottom-up procedure. In fault tree construction an outcome of concern is identified as the top event. The analyst will then explore how the top event may occur, breaking it down into contributing factors. These contributing factors will then be broken down in a cascade down the system until what are perceived to be basic events are reached.

Taking the example of a 'car failing to start' as the top event, this can be broken down into the contributing factors of 'no petrol' and 'failed ignition'. The event 'no petrol' can be broken down to 'petrol tank empty' and 'broken fuel pump'. The event 'failed ignition' can be broken down into 'flat battery' and 'failed starter motor'. The events 'petrol tank empty', 'broken fuel pump', 'flat battery', and 'failed starter motor' can be taken as the basic events.

A tree will be drawn up based on the analysis of the faults and will be composed of the events connected by 'gates'. The common gates used in practice are OR, AND, EXCLUSIVE OR, NOT, and VOTING gates. The VOTING gate assumes that there are n contributors of which r have to report failure for failure to be noted.

In an FMEA procedure one would start with the components of the system. Each component would be investigated to find its possible failure modes. Then the consequences of each failure mode would be examined for its effect on the system. For more details of this approach see O'Connor (1981) and Andrews and Moss (1993).

Both approaches suffer the same drawbacks. Analyst's perception of what may happen or what may reasonably happen guides the construction of the tree. Events which are possible may be overlooked or ignored. Frequently accidents do happen which could have been predicted and guarded against, as exemplified by the *Titanic*, *Challenger*, and many more disasters. Limiting the description of failure modes or the types of risk considered is often forced on the analysts by the size and complexity of the problem they face. There is also a reliance on the correctness of the technology or science on which the model is built. This is reinforced by cognate dissonance. Only perceived possible risks are guarded against. As a solution to these problems it is often suggested that a wide collection of experts should be consulted on the development of the tree, with hopefully a variety of backgrounds.

Having identified the causes of failure there is a need to produce an adequate and tractable mathematical description, usually the structure function. The model should provide an overall assessment and should assist in identifying components or subsystems which significantly contribute to the overall assessment. For realistic systems the problem will be very large. Sometimes analysts will attempt to circumvent this problem by dividing the system into separate modules. This must be done with extreme care since one has to ensure that the modules are independent with no interaction or interplay between the modules. It might be done occasionally if a holistic study is also undertaken.

In FTA the production of the mathematical model starts with the basic events. These are combined together as defined by the gates to produce a Boolean expression. Rapidly, as the analysis passes up the tree, the Boolean expression will become unwieldy and large, with some terms consisting of many elements. The size will reflect both the number of possible ways of failing as well as the complexity of some of the combinations which could lead to failure.

The expression is often culled to reduce its length. There are two favoured methods: culling by size and by probability. Both are carried out at the combination stage to save computation. Culling by size means the removal of any combination in which the number of basic events is above a defined threshold. It is assumed that a given number of basic events which are required to occur are unlikely to occur. The actual probability is not taken into account. In probabilistic culling combinations are removed when their probability of occurrence falls below a given threshold. Culling by size can obviously reduce the accuracy of the assessment, whilst probabilistic culling, because of the need to retain information on probabilities, is often very slow.

In most cases the FTA will produce a set of overlapping subsets composed of basic events, which are the minimal cut sets. This means that there will be a lack of uniqueness in the representation in some circumstances, such as if the tree contains a NOT gate. It will be possible to define a unique representation through the use of approaches such as consensus theory, but this will be subject to interpretation. Another problem is that for some systems, especially production processes, 'feedback' loops occur within the tree. This will cause some of the FTA algorithms to enter infinite loops.

A major drawback of the approach is that dependency is often ignored. In such cases the false assumption of independence will lead to over-optimistic estimates of the system reliability. This is particularly noted in safety systems based on redundancy, see Chapter 8 for more details. In the cases where it is not ignored the usual approach is to add in expressions for the joint failure to 'correct' the model. This seems a naive approach to the problem.

4.9 Influence diagrams

The use of an *influence diagram* is a complementary approach to the use of fault trees. Their advantages over other modelling approaches for systems are the reduction generally in the number of nodes and the explicit description of dependency within the system. They are being increasingly used in decision making and in Bayesian statistical analysis and can form a useful link between an engineer and a model builder. There are a number of texts on the subject, see Oliver and Smith (1990).

Influence diagrams are relatively simple to construct. The following approach is suggested for large diagrams:

1. Start with a preliminary list of decisions, events, and the attributes that are to be measured. Usually the attribute to be measured in reliability would be the probability of an event.

2. Identify any influences or dependencies between events and decisions

which are considered important. A clear definition of each event/decision is required for this to be easy. Give a number to all nodes and a name.

3. Order the nodes across the diagram in a time sequence.

4. Connect the nodes which influence one another with an arrow pointing in the direction of the influence.

Initially the model should be kept as simple as possible and this may be achieved by reducing the number of nodes and linkages. From the influence diagram it is possible to study the dependency in the model. If no arcs connect two events, say A, B, then they are independent. If two events can be traced back to a common event, say C, then they are jointly dependent on that event C. A and B are then conditionally dependent. If there is an arc from A to B then B is dependent on A. Studies of dependency are already seen as important in reliability and there is a fuller discussion of dependency in Chapter 8.

4.10   Discussion

This chapter has given a very basic introduction to the methods for the analysis of the reliability of a system constructed from components (subsystems). However, it is important to understand how the reliability of a system is obtained from the reliabilities of the components comprising the system. The structure function is the function which determines the system state from the states of the components (subsystems). Simple rules for the combination of structure functions allow systems to be constructed by putting subsystems in parallel or series. The information in the structure function can be summarized by the paths and cuts.

No efficient computational methods are presented for the practical evaluation of systems consisting of a large number of components as this is beyond the scope of this book. This problem has been considered by a number of authors, including Agrawal and Barlow (1984), Agrawal and Satyanarayana (1984), and Provan and Ball (1984). Because of the difficulties of exact computation of the reliability of a system consisting of a large number of components, methods for providing bounds and approximations for the reliability are discussed.

An area of research which has been of increasing interest in the last 30 years is that of the reliability of a system which is subject to two kinds of failure. It may be equally important that a system should not operate when not required as it should operate when required. Components used to construct such a system can be considered to be in one of three states. When considering the computation of the reliability of the system it is necessary to consider not only failure to operate and failure not to operate, but if the system can be in both of these states simultaneously.

5
Models for repairable systems

5.1 Introduction

Previous chapters have dealt with the distribution of lifetimes of single components and the structure of systems. In this chapter the modelling and assessment of systems whose components may be repaired is explored. These are probably the most common type of system studied in reliability engineering.

The introduction of repair brings with it a whole realm of new problems. The simplest case is if repair is instantaneous. The system would then be continuously working though it would experience events (failures). The frequency of the failures would be of interest. However, in many cases the repairs will take time and may consist of a sequence of events such as identifying the failure, sending for a repair team, the repair of a component, and the recommissioning of it.

Whilst it would be possible to devote the whole chapter to characterizing repair and the types of strategies used, only a brief outline will be given at the beginning of the chapter. At the end of the chapter optimal replacement strategies will be considered. The subject can be studied in more depth by consulting Aven and Bergman (1986), Barlow and Hunter (1960b), and G.L. Glasser (1967). The main focus, though, of this chapter will be on the description of systems behaviour through stochastic processes. Simple stochastic process models which describe the system's performance will be presented first before more complex models, which are closer to the real world, are considered. These include the modified renewal process, the alternating renewal process, the non-homogeneous Poisson process and the superimposed renewal process. Markovian models are also widely used to model many reliability systems and the analysis of these models is illustrated by methods which use differential equations. The practical application of these models is far from straightforward. There is the need to select the appropriate model and then follow the practicalities of fitting such a model. These matters will be considered in Chapter 6.

5.2   Definitions and notation

There are a number of definitions which it is necessary to introduce purely to assist in the description of repairable systems.

Repair may be defined as the attempt to correct or upgrade equipment by some method, be it mending, modifying, or adjusting the equipment. This may include corrective work on some of the components, on the structure, or on the system itself. Whilst such actions usually succeed in upgrading the equipment for a specified task, it cannot be assumed always to achieve this objective. Unfortunately it is quite a frequent experience that systems having been repaired develop even more serious faults. The most common repair strategy currently used in practice is to replace a failed or downgraded component or system with another similar component or system. Usually this will improve the system's performance, though it should not always be assumed to be the case.

Renewal of the system is usually achieved through either of the actions of repair or replacement. In a probabilistic sense the phrase is taken literally to mean to make new, and return to an initial state of performance. Only certain types of systems will have this quality. A renewal process is a process in which systems or components are renewed in a sequence. It is assumed in the process that each renewal restarts the process as new. In some cases it is assumed that the repair results in the system being returned to a level at which it was operating before failure. Such a model is referred to as a minimal repair model (Barlow and Hunter, 1960b).

Given this pattern of failure and repair (or replacement), the time to repair and the time to failure will be of interest. In reliability engineering the prime interest has been the means (expectations) of these quantities: MTTR (Mean Time To Repair) and MTTF (Mean Time To Failure). There has been some confusion over some of the terms used : MTBF (Mean Time Between Failures) is often confused with MTTF. MTBF is the length of time between a failure and the next failure, and this would include time to failure and time to repair. However, many authors assume MTBF = MTTF, rather than MTBF = MTTF + MTTR, and this shows the need to be sure that terms have been defined clearly so there is no ambiguity.

In previous chapters the system's performance could be described simply in terms of reliability, but now a system may be repaired on failure. Hence the interest is not wholly in the first time to failure, or first time to failure within a given interval, but whether a system is available and functioning, or is not functioning. *Point availability* at time t is the probability that a system is functioning at time t, which will be denoted by $A(t)$. However, the average proportion of time a system is functioning during a time period, (t_1, t_2), may be of interest. This is the *interval availability* or average availability, which will be denoted by $Av(t_1, t_2)$. It can be obtained by integrating the point availability to give

$$Av(t_1, t_2) = \frac{\int_{t_1}^{t_2} A(u)du}{t_2 - t_1} \; .$$

5.3 Repair strategies

It would be wholly inappropriate to start a chapter on repairable systems without spending a little time discussing the repair strategy itself. The definition applied in this chapter was expressed as corrective action to the system, though before such action can take place it must be decided whether it is necessary. Obviously if the system has been observed to fail then one might contemplate repair, and if a component of the system has been observed to fail then it may be replaced. This does draw attention to the concept of unrevealed failures or faults, which may exist in a system (Phillips, 1981). A system may be repaired by scheduled maintenance, or repaired after inspection. Maintenance or inspection may be at regular or variable intervals of time.

The repair of the system, or component, may be full or partial. By full repair it is implied that the system is fully explored and all parts are returned to an acceptable working state. Partial repair covers all other cases in which although some parts are upgraded others may be left in a failed state or alternatively the failures or faults are unobserved. It can also be taken to cover the cases in which repair itself occasions faults or subsequent failures (a generous interpretation).

The repair itself may be considered as *instantaneous*. This means either that the time the system is operating is the only time of interest in assessing the performance of the system and this time for repair is of no interest or that the time scale for repair is so short that it is negligible and can be ignored. Repair may take a fixed or random length of time. It may be that the repair time can be split into a number of periods. These may be the time to observation of failure, the time for diagnosis of failure, the time to organize repair resources (both labour and equipment), and then the time to carry out the repair. The repair may be carried out whilst the system is functioning or when it has failed. The former situation may occur when a component is being upgraded, which is not central to the functioning of the system.

It may be that several systems or components may fail before the system as a whole is renewed. This is the case of block repair when either there is a period until there has been a given number of failures or a given time has elapsed before proceeding to repair or replace all failed components. For example, safety systems are usually maintained on the basis of testing at regular intervals with all failures being corrected.

The wide variety of situations just described requires appropriate models, and there is unfortunately not a single comprehensive model which may be employed for such a description. In this chapter it is hoped to cover the basic models which can be adapted to fit the majority of the situations described above. However, some of the complex situations described can only be approximated by the models that will be described.

As always it is important to define the objective of the study as well as the appropriateness of the model as a description of the situation. If it is desired to estimate the number of failures of the system a fairly simple model may be employed. If, however, it is necessary to decide on the optimum maintenance strategy in detail then a more accurate description may be required.

5.4   Single component or system

Starting with models for a single component or system will give an insight to the problems while more complex systems, composed of multiple components or subsystems, can be tackled later. The models developed in this section, though, are of interest in their own right. Often interest is centred on the number of spares required to ensure a given period of operation, or on an attempt to analyse data on a sequence of components/systems.

5.4.1   RENEWAL PROCESS

Suppose initially a single component within a system is to be studied. Start by assuming that it may fail but on failure it is instantaneously repaired or replaced. The repair or replacement will bring the state of the component to an 'as good as new' state. Hence a sequence of events (failures) is observed.

The times of failures produce an ordered list of times, say $t_0, t_1, t_2, ..., t_n$, with $t_0 < t_1 < t_2 < ... < t_n$. Observation may start from the beginning of the series or from part way through, so t_0 might be zero or some starting time for observation. The times between failures are the lifetimes of the components. These lifetimes will be denoted by x_i and they will be equal to $x_i = t_i - t_{i-1}$. Fig. 5.1 represents the collection of times from the process. The distributions of the X_i are assumed to be identical and independent (IID) so that the process is a renewal process. Each repair returns the system to its original state. Such processes were of great interest in the

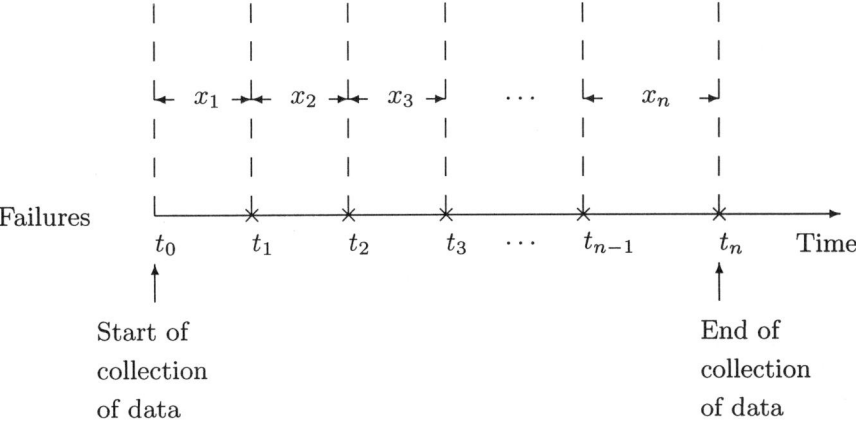

Fig. 5.1. Method of data collection for the renewal process model. × denotes an event or failure

1960s and an early text on the subject is Cox (1962).

The measures of interest might be the number of failures, the time to the jth failure, the time since the last failure, or the time to the next failure. These are helpful in deciding the number of spares required and the inspection strategy between repairs.

Assume that the underlying distribution for X_i, the component's lifetime, has a probability density function $f(x_i)$ and $t_0 = 0$. Then the measures introduced above can be expressed in terms of this function. However, before progressing to this the concept of a *convolution* must be introduced. The distribution of the time to the first failure has a probability density function, $f_1(t_1)$, which is given by $f(t_1)$, or $f(x_1)$ since $x_1 = t_1$. The distribution of the time to second failure is the distribution of the sum of the first two lifetimes, since $t_2 = x_1 + x_2$. To obtain the distribution's probability density function, $f_2(t_2)$, it is necessary to integrate over all possible combinations of x_1 and x_2 which could have led to the value t_2. This yields for $f_2(t_2)$ the convolution integral of $f(x)$ with itself given by

$$f_2(t_2) = \int_0^{t_2} f(t_2 - x_2)f(x_2)dx_2.$$

The Laplace transform of a function $f(x)$ is the integral over the positive line of the function weighted by $\exp(-sx)$, using a non-negative transform variable s. It is usually denoted by $f^*(s)$ and given as

$$f^*(s) = \int_0^\infty \exp(-sx)f(x)dx.$$

The Laplace transform is related to both the characteristic function of probability theory and the Fourier transform revered by engineers. A useful result for the Laplace transform is that the Laplace transform of a convolution of two functions is the product of the Laplace transforms of the two functions. This simplifies the expression given above since

$$f_2^*(s) = f^*(s)f^*(s) = \{f^*(s)\}^2.$$

This result can then be extended to more than two variables, so that the Laplace transform of the probability density function, $f_n(t_n)$, of t_n is given by

$$f_n^*(s) = \{f^*(s)\}^n.$$

Several tables of Laplace transforms are available, see for example Abramowitz and Stegun (1970). Hence using the tables one can find the Laplace transforms of the independent distributions and hence their product. Whilst these manipulations are eased using Laplace transforms it is still necessary to be able to invert the resulting Laplace transform to obtain the function. Again for the majority of cases tables of Laplace transforms can be used. Occasionally some manipulation may be required to express the Laplace transform in an appropriate form to use the tables. The Laplace transform itself provides useful information. It is possible to derive the expected values of the random variable by differentiation of the Laplace transform and setting the transform variable to zero, though care has to be taken with the sign which alternates. Hence for the rth moment

$$E(X^r) = (-1)^r \frac{d^r f^*(s)}{ds^r}\bigg|_{s=0}.$$

Example 5.4.1
Suppose that the distribution of times between events (failures) X has an exponential distribution with a probability density function given by

$$f(x) = \lambda \exp(-\lambda x), \text{ for } x > 0,$$

where λ is a positive constant; then the Laplace transform is

$$f^*(s) = \int_0^\infty \exp(-xs)\lambda \exp(-\lambda x)dx = \frac{\lambda}{(\lambda + s)}.$$

The Laplace transform of the probability density function of t_n, the time to the nth failure, is then given by

$$f_n^*(s) = \frac{\lambda^n}{(\lambda + s)^n}.$$

The inverse of this Laplace transform is

$$f_n(t_n) = \frac{\lambda^n t_n^{n-1} \exp(-\lambda t_n)}{(n-1)!}, \quad \text{for } t_n > 0.$$

Hence the distribution of the sum of n independent random variables with an exponential distribution with parameter λ is gamma with parameters n and λ, as was stated in Chapter 2, Subsection 2.3.1.

□

So far the time to the nth failure has been considered, and now one of the other measures of interest for renewal processes will be considered. This is the cumulative number of failures (events) which occur for a process in the period $(0, t)$, which will be denoted by $N(t)$. The distribution of $N(t)$ will be discrete and is defined on the non-negative integers, 0, 1, 2 .....

If $N(t) = k$ then there must have been k failures in the period $(0, t)$ and the $(k+1)$th failure must occur after the time t. Hence t_k occurs in the period up to t and there has not been an event in the period t_k to t. The process survived the period t_k to t without a failure. Hence the probability can be obtained from the convolution of $f_k(t_k)$ and $R(t)$, the reliability function of the distribution of the lifetimes, and is given by

$$P(N(t) = k) = \int_0^t f_k(t - y)R(y)dy.$$

Hence at any point of time the distribution of the number of failures can be obtained. Since the number of spares required may be of interest, it will be useful to obtain the expectation of the number of failures in the period $(0, t)$, which is given by

$$E(N(t)) = \sum_{k=0}^{\infty} kP(N(t) = k) = \sum_{k=0}^{\infty} k \int_0^t f_k(t - y)R(y)dy.$$

$E(N(t))$ is usually referred to as the *renewal function* and denoted by $H(t)$. Again for most purposes it is easier to use the Laplace transform of the renewal function, which is given by

$$H^*(s) = \sum_{k=0}^{\infty} kP^*(N(t) = k) = \sum_{k=0}^{\infty} k\{f^*(s)\}^k R^*(s).$$

$R^*(s)$ can be written as $(1 - f^*(s))/\mu$, where $\mu \ (= E(X))$ is the expectation of the distribution of the lifetimes, and hence $H^*(s)$ is given by

$$H^*(s) = \frac{f^*(s)}{\mu(1 - f^*(s))}.$$

From this result it is possible to derive the approximate renewal function as t increases towards infinity. Under fairly general conditions the renewal function can be expressed, as t tends to infinity, as

$$H(t) = \frac{t}{\mu} + \frac{\sigma^2 - \mu^2}{2\mu^2},$$

where σ^2 is the variance of the lifetime X. For a proof of this result see Beaumont (1983), Ross (1970), or Cox (1962). Often interest centres on the first term, especially when evaluating infinite time horizon repair/replacement strategies, see Barlow and Hunter (1960b). However, other authors have considered the inclusion of the second term, see Ansell et al. (1984).

Whilst $H(t)$ gives the expected number of failures in the period $(0, t)$ for the process, sometimes it is more useful to measure the rate of change, which can be obtained by differentiation. The function obtained is denoted by $h(t)$ and is known as the *renewal density function* or ROCOF (Rate of OCurrence Of Failures). Then the probability of a failure of the process in the next small interval $(t, t + dt)$ is given approximately by $h(t)dt$. This function should **not** be confused with a component's hazard rate. The renewal density function reflects the size of the probability for the next failure in the system at time t, which may be the first, second, or kth. The hazard function reflects the size of the probability of failure for a single component given that it has worked until time t.

5.4.2 HOMOGENEOUS POISSON PROCESS (HPP)

A special case of the renewal process is the homogeneous Poisson process. If the distribution of time between events (failures) is assumed to be an exponential distribution, as in Example 5.4.1, and if each component operates independently then the process is known as the HPP (Homogeneous Poisson Process). This implies that the distributions of the number of failures in non-overlapping intervals of time are independent. The assumption that the lifetime distribution is exponential also implies that in a small interval of length dt the probability of a failure occurring is λdt, for some positive constant λ, and the probability of more than one failure occurring is $o(dt)$, (which means that $o(dt)/dt \rightarrow 0$ as $dt \rightarrow 0$), see Chapter 1, Example 1.6.3, and Chapter 2, Subsection 2.3.1. In fact if any one of the statements in the last three sentences is true then this implies that the statements in the other two sentences are true. An axiomatic description of the process is given in many texts such as Beaumont (1983), Ascher and Feingold (1984), and Cox (1962).

If $f(x) = \lambda \exp(-\lambda x)$ then from Example 5.4.1 the probability density function of t_k, the time to the kth failure, is given by

$$f_k(t_k) = \frac{\lambda^k t_k^{k-1} \exp(-\lambda t_k)}{(k-1)!}$$

and the distribution of the number of failures in period $(0, t)$ is given by

$$P(N(t) = k) = \int_0^t \frac{\lambda^k (t - t_k)^{k-1} \exp(-\lambda t)}{(k-1)!} dt_k = \frac{\lambda^k t^k \exp(-\lambda t)}{k!},$$

which is the Poisson distribution with parameter λt. Hence the name of the process as the distribution of the number of failures in the interval (t_1, t_2) is also Poisson with parameter $\lambda(t_2 - t_1)$. So the 'failure' rate is constant (homogeneous) and equal to λ. This is confirmed by considering the renewal function, which is given by

$$H(t) = \sum_{k=0}^{\infty} k P(N(t) = k) = \sum_{k=0}^{\infty} k \frac{\lambda^k t^k \exp(-\lambda t)}{k!} = \lambda t,$$

and the renewal density function (ROCOF), which is given by

$$h(t) = \lambda.$$

The process is characterized by a 'lack of memory' in that it does not take account of the time since the last event. This is why there is frequent confusion between the *hazard function* and the *renewal density function*, since for this process they are the same (λ). It is the model quite frequently, though often inappropriately, used to described failures in reliability. The model is frequently used for electronic data where the assumption of random failures in time is realistic. It can also arise from more complex processes, at least as a first approximation, see Subsection 5.5.3.

5.4.3 MODIFIED RENEWAL PROCESS

If t_0 is not the start of the process or if the distribution of the time to the first failure differs from the distributions of the times to failure of the second, third, or subsequent components, then the process is a *modified renewal process*. A special case of this is the *equilibrium renewal* process. In this case observation starts after the process has been running for some time and thus the system may have reached a state of equilibrium. Both of these types of processes have been studied in detail, see Cox (1962).

In the case of modified processes, whose first failure distribution differs from the rest, then the above results can be simply amended using this different first distribution. Hence if $f_0(t)$ is the probability density function of the first component/system then the time to the kth failure would be given by the convolution

$$f_k(t) = \int_0^t f_0(t-u)f_{k-1}(u)du$$

and similarly its Laplace transform is

$$f_k^*(s) = f_0^*(s)f_{k-1}^*(s).$$

Other quantities of interest can be derived similarly. In the case of the equilibrium distribution, where observation of the process starts at some time after the process begins, the first probability density function is replaced by $R(t)/\mu$, the reliability function divided by μ, the mean of the lifetime distribution. This will again amend all the previous results accordingly.

5.4.4   ALTERNATING RENEWAL PROCESS

The renewal processes described in the previous sections are rather simple in assuming that repair or even replacement is 'instantaneous', and so take either no time or a negligible amount of time. If the repair times are significantly different from zero then it is necessary to extend the model. Hence the ith failure at time t_i is followed by a repair time which ends at time r_i, as is illustrated in Fig. 5.2. This obviously produces a series of alternating states with the component either working or failing. A simple model of this is the *alternating renewal process*. In such a process the distributions of the series of failure times, x_i, are assumed to be identical as are the distributions of the repair times, y_i. It is also assumed that the failure and repair times are independent. There is a well-developed theory again for the alternating renewal process, see Cox (1962). The distribution of time to the first failure will be, as before, simply the distribution of the first lifetime, X_1. The time to the second failure at time t_2 will be composed of the time, x_1, to the first failure at time t_1, the length of time of the first repair, y_1, and the time, x_2, from the end of that repair at time r_1 until the time of the second failure at time x_2. Hence the probability density function of T_2, the time to the second failure, will be given by the convolution

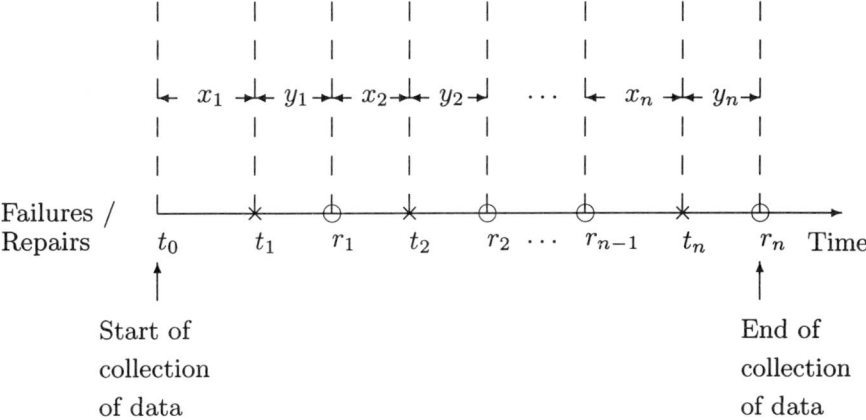

Fig. 5.2. Method of data collection for the alternating renewal process model. × denotes an event (failure) and ⊖ denotes the end of a repair time

$$f_2(t_2) = \int_0^{t_2} \int_0^{r_1} f(t_2 - r_1)g(r_1 - t_1)f(t_1)dt_1 dr_1,$$

where $g(y_1)$ is the probability density function of the length of time of the repair. Again it is probably easier to express this in terms of Laplace transforms, so that

$$f_2^*(s) = f^*(s)g^*(s)f^*(s).$$

Hence by repeated application of this method the Laplace transform of the probability density function of t_n, the time to the nth failure, is given by

$$f_n^*(s) = f^*(s)\{g^*(s)f^*(s)\}^{n-1}.$$

Given that the repair time is not negligible then the system or component will not be functioning at certain periods and hence will be unavailable. It is therefore sensible to consider the availability of a component, either the point availability, $A(t)$, or the average interval availability, $Av(t_i, t_j)$, over the interval (t_i, t_j). The point availability is the probability the component is working at time t, which is given by

$$A(t) = \sum_{k=1}^{\infty} \int_0^t f_k(t - u)R(u)du,$$

where $R(t)$ is the reliability function of the distribution of the failure time of a component.

Then $Av(t) = Av(0, t)$ can be computed from $A(t)$ and is given by

$$Av(t) = \frac{\int_0^t A(u)du}{t}.$$

The long run equivalents of these are easier to obtain for this process. Since it is assumed that the distributions are identical and independent and the mean time to failure, MTTF, and the mean time to repair, MTTR, are finite, the asymptotic availability is given by

$$As = Av(\infty) = \frac{\text{MTTF}}{\text{MTTF} + \text{MTTR}}.$$

This is a property of the stationarity of the process. Note that this result is often incorrectly misquoted as applying to all stochastic models, see Ascher and Feingold (1984).

Whilst it is possible to regard this model as very simplistic for systems behaviour it forms the underlying description for most component behaviour. It is possible to extend the concept of alternating renewal processes to cover other aspects such as waiting for detection, diagnosis, etc. The level of complexity can be increased, though it should be remembered before considering such complexity that the modelling is being undertaken to achieve an objective.

5.4.5   NON-HOMOGENEOUS POISSON PROCESS (NHPP)

Assuming that the components or systems are identical is a very restrictive assumption. It is also the case that the components may be affected by changes in the environment or there may be an upgrading of the components. (It is also possible that there is a downgrading of the components.) Under such circumstances the distributions of lifetimes $x_1, x_2, ..., x_n$ cannot be assumed to be identical and other models have to be contemplated.

The worst case would be that every component lifetime was taken from a different underlying distribution. This would allow little possibility of analysis. Fortunately the component lifetimes are usually assumed to come from the same family of distributions and only vary through the parameters.

An often used extension to the HPP is the NHPP (Non-Homogeneous Poisson Process). The rate of failures, as given by the renewal density function (or ROCOF) $h(t)$ at time t, is assumed to vary with time. In the context of the NHPP this function is also referred to as the *peril rate*. A number of authors consider an axiomatic definition, generalizing a single axiom of the homogeneous case, see Ascher and Feingold (1984). The NHPP occurs frequently as a special case of more complex models.

Example 5.4.2

Two particular models for the NHPP have been studied in detail and the renewal density functions (ROCOFs) are now given.

(a) Model I (*Crow's model*): $h(t) = \lambda\beta t^{\beta-1}$, where λ and $\beta > 0$.

Model I is sometimes referred to misleadingly as the Weibull rate model. This model is often used to model reliability growth, see Chapter 7, Subsection 7.2.4. λ is the scale parameter and β is the growth parameter. For $\beta < 1$ the process will improve with time and for $\beta > 1$ it will deteriorate. Model I has been considered by several authors, including Crow (1982).

(b) Model II (*Cox and Lewis's model*): $h(t) = \exp(a_0 + a_1 t)$.

Model II has the advantage over model I in that the parameters are not restricted and hence estimation can be easier. So a_0 is like the scale parameter in this case and a_1 will indicate growth. Model II is discussed in detail in Cox and Lewis (1966) and has been extended beyond the linear form to other polynomials, see Lewis (1972) and Lewis and Shedler (1976).

□

As for the HPP the number of failures in an interval for the NHPP is Poisson, though the expectation is the integral of the renewal density function (peril rate). Hence for an interval (t_1, t_2) with peril rate $h(t)$ the expectation for the Poisson distribution is

$$H(t_2) - H(t_1) = \int_{t_1}^{t_2} h(u)du.$$

5.5 Multiple components

In moving to study multiple components there is a need to decide whether interest is centred primarily on the components or the overall performance of the systems in which they are contained. It is possible that sole interest may be in the number of spares to ensure that every failed component can be repaired. Alternatively the reliability of the whole system may be of interest. In this case it is necessary to know the number of spares required to ensure that the system reaches a given level of reliability.

5.5.1 COMPONENT INTEREST

If a system consists of n components, each of which on failure may be repaired or replaced, then there are two models which are generalizations of the renewal process model: the branching process and the SRP (Superimposition of Renewal processes or Superimposed Renewal process).

5.5.2 BRANCHING PROCESS

The branching process model assumes that there are a set of initiating events, referred to as the primary events, which follow either an HPP or NHPP. These primary events give rise to subsidiary events. For any primary event there will be a random number, say s, of possible subsidiary events, including the possibility of $s = 0$. The s events then form a renewal process with an assumed known distribution. Obviously a special case of this process is when the s events occur simultaneously together. However, the procedure for estimating the parameters of this model, as described by Cox and Lewis (1966), seems a little discouraging to anyone applying the model to reliability data. Lewis (1964) showed for the homogeneous case that it can model very adequately failure data in a complex situation, with an application to hardware failures in computing equipment. Further details of the non-homogeneous case are given in Lewis (1967).

5.5.3 SUPERIMPOSITION OF RENEWAL PROCESSES (SRP)

Again starting with the simplest model assume that repair is instantaneous and that m independent renewal processes are being observed. If the time to the next event (failure) or the number of failures is of interest then the 'sum' of these m sequences of the renewal processes is required. This is referred to as a SRP and is illustrated in Fig. 5.3. The overall effect of the m sequences is given at the bottom of Fig. 5.3. Results for this process are given in Cox (1962).

Example 5.5.1

The simplest case is the superimposition of m HPPs with a renewal density of λ_i for the ith process. The superimposition in this case will give an HPP with a renewal density function of $\sum_{i=0}^{m} \lambda_i$.

□

The superimposition is a relatively complex process and for most non-trivial cases the distribution of time to the next event is far from simple. Hence there is interest in whether it is possible to describe the resulting pattern of failures given by the superimposition of renewal processes by simpler processes. Several authors have considered this problem.

For two independent renewal processes to result in a renewal process on superimposition Cinlar (1975) showed that all the processes must be HPPs. Hence, in general, it is not possible that superimpositions give renewal processes. However, asymptotic results exist which suggest under certain circumstances that it is possible to approximate the superimposition by simpler models. Drenick (1960), Khintchine (1960), and Cox (1962) showed that the superimposition of equilibrium renewal processes is asymptotically an HPP as the number of renewal processes increases. Further Cox

Component

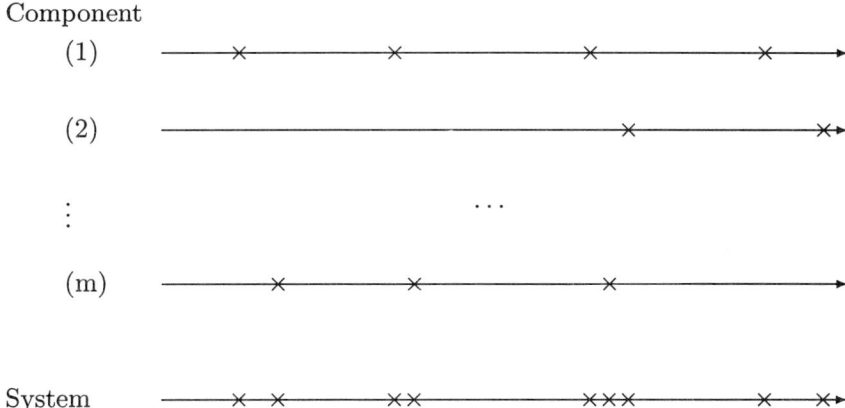

Fig. 5.3. Superimposition of renewal processes. × denotes an event or failure

(1962) pointed out two consequences of this result: firstly that this gives an explanation in practice of processes being closely approximated by the HPP; and secondly that the analysis of the superimposition will give little information about the individual underlying processes. Grigelionis (1964) gave limiting results for ordinary renewal processes, the transient behaviour of which Blumenthal *et al.* (1973) considered. From these results it is seen that the superimposition may behave like an NHPP when there are a large number of processes.

This implies that for some large systems it may be possible to represent the number of failures at least by an NHPP or even an HPP. This point will be covered in more detail in Chapter 6. Again it is possible to extend this model by assuming that each sequence is an alternating renewal process, or even a more general process. There may be interaction between the sequences, and this dependency is covered in Chapter 8.

5.5.4 SYSTEM INTEREST

If the main concern is the performance of the system then it may be that the repair times are significantly different from zero. Hence there will be intervals when components will not be functioning. Assuming that the system's performance is defined by the components' performance through a structure function, as described in Chapter 4, then the system will either be working or have failed. It will be possible to define the set of states for

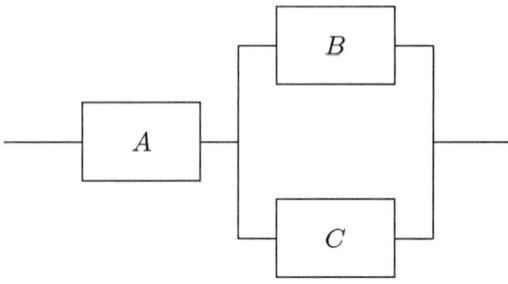

Fig. 5.4.   Three-component system

the components which will result in the system working and a set for which the system will have failed. At the system level there will be interest in the *state probability*, the probability of the system being in a given state at time t. The usual approach to modelling such systems in reliability studies is by using differential equations or simulation.

5.5.5   DIFFERENTIAL EQUATIONS

The modelling can be relatively simple as will be shown. The solution of the differential equations developed even in the simplest case of exponential failure and repair times can be difficult. It is fortunate that for most cases the equations derived are well behaved and lead to solutions. Complexity arises mainly from the size of the problem considered. It has therefore become common practice to consider only steady state results rather than results for the transitory behaviour between the initial state and the steady state. There exists a large literature on the use of differential equations for stochastic modelling in reliability studies; see Billington and Allan (1983) and Cox and Miller (1965) for further details.

5.5.6   MARKOVIAN MODELS

Stochastic modelling through the use of differential equations developed both in the modelling of life systems and queues. The terminology such as birth, death, immigration, and emigration comes from life systems. The

Table 5.1. System states of three-component system

Component	Component state A	B	C	System state
	1	1	1	1
	1	1	0	1
	1	0	1	1
	0	1	1	0
	1	0	0	0
	0	1	0	0
	0	0	1	0
	0	0	0	0

HPP is an example of a simple pure birth process. Now a model composed of components which may pass from working to failed and failed to working states will be considered, and the state probabilities obtained.

It is assumed that in a small time period of length dt the chance of more than one event occurring is negligible if the components are functioning independently. Strictly the probability of more than one event occurring is $o(dt)$, which means that $o(dt)/dt \to 0$ as $dt \to 0$. Hence it is only necessary to consider either a failure or a repair but not both in this small interval. Therefore by the end of the interval either a component may fail, which may cause the system to fail, or a component may be repaired, which may cause the system to work or nothing may happen. In the Markovian model it is assumed that in an interval of given length the probabilities of a working component failing (or a failed component being repaired) are only dependent on the state of the system at the beginning of this small interval and the length of the interval. This 'lack of memory' property means that information on how the system entered the current state or when it entered the state is not needed.

Example 5.5.2

As outlined in Chapter 4, if the system depends on the state of components then it is possible to define the set of states of components for which the system is working and for which it is failed. Fig. 5.4 shows a system which works provided component A works **and** B and/or C works. The states of the components are given in Table 5.1 as well as the state of the system.

In this example the system only works in the first three states: when A, B, and C work; A and B work; and A and C work. In all other states the system fails.

Assuming the Markovian model the failure rates in an interval $(t, t+\delta t)$ for the components A, B, and C, when they are working, are λ_a, λ_b, and λ_c, respectively, and the repair rates, when they are failed, are μ_a, μ_b, and μ_c, respectively.

Suppose at the end of the interval the system is functioning with A and B working and C being repaired. Then since the beginning of the interval either nothing has happened and A and B were working and C was being repaired at the beginning, or A was repaired and A and C were being repaired with B working, or B was repaired and A was working with B and C being repaired, or C has failed and all components were working. Otherwise multiple events would have occurred. This can be expressed in the equation

$$
\begin{aligned}
P(A, B; t + dt) \quad = \quad & (1 - \lambda_a dt - \lambda_b dt - \mu_c dt) P(A, B; t) + \mu_a dt P(B; t) \\
& + \mu_b dt P(A; t) + \lambda_c dt P(A, B, C; t) + o(dt),
\end{aligned}
$$

where $P(A, B; t)$ is the probability that A and B are working and C has failed at time t; $P(B; t)$ is the probability that B is working and A and C have failed at time t; $P(A; t)$ is the probability that A is working and B and C have failed at time t; and $P(A, B, C; t)$ is the probability that A, B, and C are working at time t. The equation can be rewritten as

$$
\begin{aligned}
P(A, B; t + dt) - P(A, B; t) \quad = \quad & -(\lambda_a dt + \lambda_b dt + \mu_c dt) P(A, B; t) \\
& + \mu_a dt P(B; t) + \mu_b dt P(A; t) \\
& + \lambda_c dt P(A, B, C; t) + o(dt).
\end{aligned}
$$

Dividing both sides by dt and letting $dt \to 0$ then gives

$$
\begin{aligned}
P'(A, B; t) \quad = \quad & -(\lambda_a + \lambda_b + \mu_c) P(A, B; t) + \mu_a P(B; t) \\
& + \mu_b P(A; t) + \lambda_c P(A, B, C; t),
\end{aligned}
$$

where $P'(A, B; t)$ is the derivative of $P(A, B; t)$ with respect to t. Hence a differential equation for $P(A, B; t)$ is obtained. Other equations can be derived in a similar fashion. For completeness all the equations are presented here, extending the $P(\quad ; t)$ notation in the natural way and using $P(0; t)$ as the probability of no components working at time t:

$$
\begin{aligned}
P'(0; t) \quad = \quad & -(\mu_a + \mu_b + \mu_c) P(0; t) + \lambda_a P(A; t) \\
& + \lambda_b P(B; t) + \lambda_c P(C; t)
\end{aligned}
$$

$$
\begin{aligned}
P'(A;t) &= \mu_a P(0;t) - (\lambda_a + \mu_b + \mu_c)P(A;t) \\
&\quad + \lambda_c P(A,C;t) + \lambda_b P(A,B;t) \\
P'(B;t) &= \mu_b P(0;t) - (\mu_a + \lambda_b + \mu_c)P(B;t) \\
&\quad + \lambda_c P(B,C;t) + \lambda_a P(A,B;t) \\
P'(C;t) &= \mu_c P(0;t) - (\mu_a + \mu_b + \lambda_c)P(C;t) \\
&\quad + \lambda_b P(B,C;t) + \lambda_a P(A,C;t) \\
P'(B,C;t) &= \mu_c P(B;t) + \mu_b P(C;t) \\
&\quad - (\mu_a + \lambda_b + \lambda_c)P(B,C;t) + \lambda_a P(A,B,C;t) \\
P'(A,C;t) &= \mu_c P(A) + \mu_a P(C;t) \\
&\quad - (\lambda_a + \mu_b + \lambda_c)P(A,C;t) + \lambda_b P(A,B,C;t) \\
P'(A,B;t) &= \mu_b P(A;t) + \mu_a P(B;t) \\
&\quad - (\lambda_a + \lambda_b + \mu_c)P(A,B;t) + \lambda_c P(A,B,C;t) \\
P'(A,B,C;t) &= \mu_a P(B,C;t) + \mu_b P(A,C;t) + \mu_c P(A,B;t) \\
&\quad - (\lambda_a + \lambda_b + \lambda_c)P(A,B,C;t).
\end{aligned}
$$

This set of equations can be expressed in matrix form as

$$\mathbf{P}'(t) = \mathbf{P}(t)\mathbf{Q},$$

where

$$
\begin{aligned}
\mathbf{P}(t) &= (P(0;t), P(A;t), P(B;t), P(C;t), \\
&\qquad P(B,C;t), P(A,C;t), P(A,B;t), P(A,B,C;t)),
\end{aligned}
$$

$\mathbf{P}'(t)$ is a vector of the derivatives of $\mathbf{P}(t)$, and $\mathbf{Q}$ is a matrix of transition (failure or repair) rates, whose rows sum to zero.

□

This example illustrates the result that the vector of state probabilities $\mathbf{P}(t)$ is a solution of the differential equations, which are given in matrix form by

$$\mathbf{P}'(t) = \mathbf{P}(t)\mathbf{Q},$$

where $\mathbf{Q}$ is a matrix of transition (failure or repair) rates, whose rows sum to zero. There are a number of methods for solving these equations using Laplace transforms, generating functions, or by matrix methods, see for example Beaumont (1983), Cox and Miller (1965), and O'Connor (1981). The solutions produced will give the transitory behaviour of the system which will be a function of time and be dependent on the initial conditions. In some analyses the transitory behaviour is important, for example power generation models considered by Billington and Allan (1983).

Example 5.5.3

Consider the two-component parallel system, which works provided one out of the two components works, (1-out-of-2 system). If the failure rate is λ and the repair rate is μ then the system of equations can be obtained from Example 5.5.2 by using components B and C and excluding component A using rates $\lambda_a = 0$, $\lambda_b = \lambda_c = \lambda$, $\mu_a - 0$, and $\mu_b = \mu_c - \mu$. Then defining $P_2(t) = P(B, C; t)$, the probability of two components working, $P_1(t) = P(B; t) + P(C; t)$, the probability of one component working, and $P_0(t) = P(0; t)$, the probability of no components working at time t, the system of differential equations can be written as follows:

$$
\begin{aligned}
P_0'(t) &= -2\mu P_0(t) + \lambda P_1(t) \\
P_1'(t) &= 2\mu P_0(t) - (\lambda + \mu)P_1(t) + 2\lambda P_2(t) \\
P_2'(t) &= \mu P_1(t) - 2\lambda P_2(t).
\end{aligned}
$$

In matrix form this would be

$$
\mathbf{P}'(t) = \mathbf{P}(t) \begin{pmatrix} -2\mu & 2\mu & 0 \\ \lambda & -(\lambda + \mu) & \mu \\ 0 & 2\lambda & -2\lambda \end{pmatrix},
$$

where $\mathbf{P}'(t) = (P_0'(t), P_1'(t), P_2'(t))$ and $\mathbf{P}(t) = (P_0(t), P_1(t), P_2(t))$. Hence

$$
\mathbf{Q} = \begin{pmatrix} -2\mu & 2\mu & 0 \\ \lambda & -(\lambda + \mu) & \mu \\ 0 & 2\lambda & -2\lambda \end{pmatrix}.
$$

It is possible to eliminate $P_1(t)$ from the third equation and obtain a second-order differential equation for $P_2(t)$, which when solved gives

$$
P_2(t) = \frac{\mu^2 + 2\lambda\mu \exp(-(\lambda + \mu)t) + \lambda^2 \exp(-2(\lambda + \mu)t)}{(\lambda + \mu)^2},
$$

assuming that the two components are working at $t = 0$. This result can then be substituted in the third equation to give

$$
P_1(t) = \frac{2\lambda\mu + 2\lambda(\lambda - \mu) \exp(-(\lambda + \mu)t) - 2\lambda^2 \exp(-2(\lambda + \mu)t)}{(\lambda + \mu)^2}
$$

and hence

$$
P_0(t) = \frac{\lambda^2 - 2\lambda^2 \exp(-(\lambda + \mu)t) + \lambda^2 \exp(-2(\lambda + \mu)t)}{(\lambda + \mu)^2}
$$

can be obtained using the second equation. To illustrate their transitory behaviour these three probabilities are plotted in Fig. 5.5 for the case when

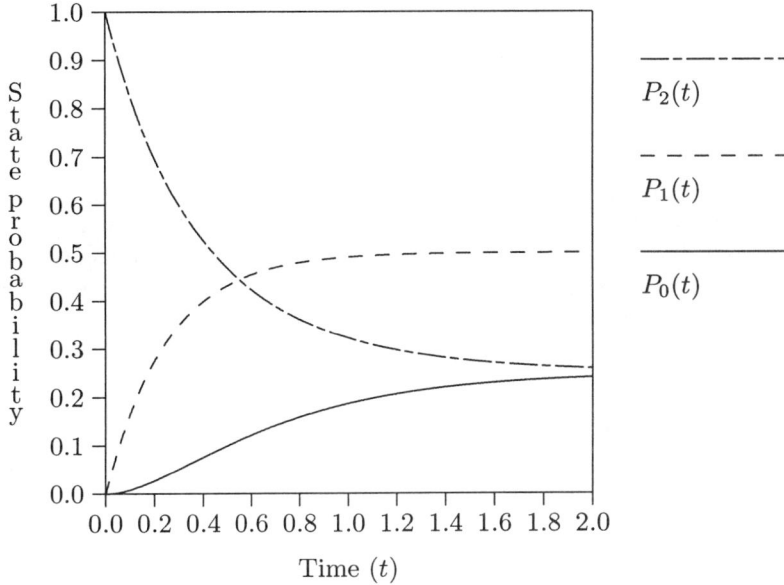

Fig. 5.5. State probabilities for a two-component parallel system if $\lambda = \mu = 1$

$\lambda = \mu = 1$. From this plot it is seen that the probabilities rapidly approach their limiting (asymptotic) values as t becomes greater than 1.

The system (point) availability is given by

$$A(t) = 1 - P_0(t)$$

and hence decreases monotonically from 1 to 0.75, when $\lambda = \mu = 1$.

□

However, in most reliability contexts, given the size of the problem, the use of the transitory behaviour will be too sophisticated an approach. Also if there is a lack of information on component behaviour, this will justify not considering the transitory behaviour. In some cases it may well not be possible through the lack of computing resources to consider the transitory behaviour. Most reliability analysts will therefore consider the asymptotic behaviour of the system. It should be stressed, however, that by this approach an approximation is being used which may not reflect behaviour well especially in the initial period.

5.5.7   STEADY STATE SOLUTIONS

The steady state results can be obtained by taking the transitory solutions and letting t tend to infinity. Alternatively for most systems these results can be obtained by setting $\mathbf{P}'(\mathbf{t}) = \mathbf{0}$ in the matrix equations in Example 5.5.3, which results in the equations

$$0 = \mathbf{PQ},$$

where $\mathbf{P} = \mathbf{P}(\infty)$. These can be readily solved in most cases encountered in reliability.

Example 5.5.4

Consider the two-component parallel system, which works provided one out of the two components works, (1-out-of-2 system), which was considered in Example 5.5.3. The set of differential equations for the state probabilities are given in matrix form as

$$\mathbf{P}'(t) = \mathbf{P}(t)\mathbf{Q}$$

where $\mathbf{P}'(t) = (P_0'(t), P_1'(t), P_2'(t))$ and $\mathbf{P}(t) = (P_0(t), P_1(t), P_2(t))$ and

$$\mathbf{Q} = \begin{pmatrix} -2\mu & 2\mu & 0 \\ \lambda & -(\lambda+\mu) & \mu \\ 0 & 2\lambda & -2\lambda \end{pmatrix}.$$

In the steady state the equations become

$$(0,0,0) = (P_0, P_1, P_2) \begin{pmatrix} -2\mu & 2\mu & 0 \\ \lambda & -(\lambda+\mu) & \mu \\ 0 & 2\lambda & -2\lambda \end{pmatrix},$$

where $P_i = P_i(\infty)$, for $i = 1, 2, 3$. The solution to these equations is

$$P_0 = \lambda^2/(\mu+\lambda)^2, P_1 = 2\mu\lambda/(\mu+\lambda)^2, P_2 = \mu^2/(\mu+\lambda)^2.$$

If $\lambda = \mu = 1$ these asymptotic probabilities become

$$P_0 = 0.25, P_1 = 0.5, P_2 = 0.25,$$

which are the limits obtained in Fig. 5.5.

□

5.5.8   SEMI-MARKOVIAN MODELS

In the above model it is assumed that $\mathbf{Q}$ was not dependent on time. It is possible to generalize the model to consider examples where $\mathbf{Q}(t)$ does depend on time, t. This implies that the time in a state is not exponential and a wide range of possible distributions could be considered. The model produced is the semi-Markovian model, see Cox and Miller (1965). Again there are a variety of ways of tackling such models, such as by the use of supplementary variables. However, this is judged to be outside the scope of this book.

5.5.9   SIMULATION

An alternative approach to the solution of the differential equations for the state probabilities is to use Monte Carlo simulation. It is possible to employ this method when solutions by other methods do not appear tractable. The level of sophistication in the simulation will depend on the objective of the study. There are a number of simulation packages available which offer a wide range of facilities such as a variety of distributions for failure and repair times. A thorough text on simulation in statistics is Ripley (1977).

The main problem encountered with simulation is the credence that can be given to the results. Since in reliability studies components are generally thought of as very reliable and repair times are assumed short then a large number of component events need to be simulated before a system failure is encountered. The lack of data means that it is difficult to assess the precision of the results with any degree of confidence. There are some methods of designing simulation experiments which may increase precision, see Ripley (1977). A sensible strategy, though, in using simulation is to use it in conjunction with steady state results by simulating a simple system and comparing the results to the steady state results.

5.6   Optimal replacement

The aim of this section is to give again a flavour of an extensive field. Those who require more detail should consult Osaki (1985), Sherif and Smith (1981), or Thomas (1986).

To illustrate the approach the case of optimal replacement of a component in a sequence of components will be considered. Each component can be replaced on failure or alternatively replaced at a fixed time t'. It is assumed that the cost of allowing a component to fail and then be replaced, c_1, is higher than preventative replacement at time t', c_2. In most practical cases a finite horizon is assumed, say T, beyond which either the system will not be required or the accounting period is finished. The objective is therefore to choose t' such that the expected cost is minimized. The expected cost EC for a given t' will be given by

$$EC = \sum_{i=1}^{N(T)} E(C_i),$$

where $E(C_i)$ is the expectation of C_i, the cost of replacement of the ith item whose lifetime was t_i (where C_i will either be c_1 if $t_i < t'$ or c_2 if $t_i > t'$) and $N(T)$ is the number of failures in $(0, T)$ under the policy. If t' is large there will be a high number of costly failures but the number of events will be lower given the longer expected life of the components. If t' is small then there will be more replacement but at a lower cost. It is the number of events which is being traded off against their costs.

It is possible using dynamic programming to obtain the optimal replacement time t' for the finite time problem, see Barlow and Proschan (1965), but the procedure has been considered computationally long winded with the need for numerical integration and search methods for even simple cases such as failure times with the gamma distribution. For these reasons interest has centred on asymptotic approximations. The usual approach is to select t' to minimize the asymptotic average cost per unit time which is equivalent to the expected cost in a replacement divided by the expected length of time to replacement. This equivalence is due to Wald's theorem and the approximation to the renewal function given earlier in Subsection 5.4.1. Hence the optimal t' is obtained by minimizing $E(C_i)/E(T_i)$. Christer (1978) proposed taking further terms in the expansion of the renewal function to obtain a better solution for large but finite T. Ansell et al. (1984) showed that Christer's proposal offered little advantage.

Others have considered models in which t' is not fixed but depends on the time remaining until the time horizon is reached. Again this requires dynamic programming. Other authors, for example Jack and Dagpunar (1992), have assumed that the underlying process is not a renewal process but an NHPP. The argument for doing this is that repairs only return the system back to the state just before failure, which is a minimal repair model.

Increasingly interest has centred on information which may be available either directly on the process or from covariates. These models allow for inspection and the decision to replace to be made on the inspection information, see Jardine and Anderson (1984) and Makis and Jardine (1992). These authors have made use of the proportional hazards model to estimate the probability of survival of components.

5.7   Discussion

This chapter has considered the main stochastic models used for repairable systems. It has considered the main results which are necessary for models which will be used later in the text. It has also attempted to cover briefly the use of differential equations in reliability studies and the problems associated with optimal replacement. The fitting and assessment of the models are covered in Chapter 6.

6
Analysis of repairable systems

6.1 Introduction

This chapter will explore the analysis of data from repairable systems using the models which were discussed in Chapter 5. Knowledge about the process will affect the analysis and this knowledge may have established what is the nature of the underlying process. However, this will not always be the case and sometimes all that will be available will be a list of failure times and possibly also repair times. In some cases this may be aided by knowledge of the failure modes and possibly some knowledge of the system. Obviously the more information available then the better the level of modelling should be. Cox and Lewis (1966) is a good introduction to this area and Ascher and Feingold (1984) is also a useful text on this topic. This chapter concentrates on fitting a variety of models to repairable system data and starts with a few brief comments on the fitting (estimating the parameters) of the models and the assessment of fit. For a more detailed treatment Cox and Snell (1971) is a useful starting point.

As with any analysis it is useful to start with simple models and build to more complex models, as was done in Chapter 5. Hence analysts often start by assuming the data may come from a homogeneous Poisson process (HPP) and only reject such a model when it is evident that it is inappropriate. An HPP is a renewal process in which the inter-arrival times have an exponential distribution with the same renewal rate throughout the observed period. This simple model may be inappropriate for many reasons: the underlying distribution may not be exponential, there may be a change in the rate over time, components may be dependent, or the process may not be stationary. Each of these cases will be considered in this chapter as well as other aspects such as the prediction of the future performance of the system. Non-parametric estimators may be a useful guide to decide which model to fit and so estimators of the renewal density function (ROCOF) and the renewal function are presented. It will be assumed in this chapter that the data are the chronologically ordered times taken throughout the observation period at the times the system was repaired.

6.2   Fitting models to data

Given the number of possible models which could be applied to repairable systems some guidance on choice of model is necessary. There are two important aspects of choosing a model. Firstly, does the model provide understanding of the system which appears to be plausible to the analyst? Secondly, does the model give a reasonable fit to the data? A model should always help to bring insight to the analysis. Simply choosing the model which fits the data 'best' is a dubious approach. There does need to be a rationale for the fitting of the model. A good fit to historic data does not necessarily imply a good predictor of future data. A poorer fit which provides comprehension of the data will often provide better predictive capability. Hence the selection of the 'best' model depends on the context being studied and the physical interpretation that can be given to the model. Given that more than one model seems plausible, selection should be based on the model which provides an adequate fit to the data. There is a great danger of fitting an over-elaborate model when a simple model provides nearly the same level of explanation. Parsimonious fitting is commended, so that simpler models should be fitted if they provide an almost equivalent fit to the data as more complex models. A model with few parameters is preferred to a model with many, unless the latter has considerable advantage in fitting the data better.

When fitting data there are three stages: estimating unknown parameters, testing their significance, and assessing the fit. Estimation can be carried out by the methods of maximum likelihood, least squares, or a Bayesian approach as suggested in Chapter 2. Hypothesis tests can be based on either exact or approximate distributional knowledge. Assessment of the fit of the model is more complex. It may be judged using 'goodness-of-fit' tests to assess how well the data are explained by the model. An associated aspect is whether there are influential observations or outliers in the data, see Cook and Weisberg (1982). An influential observation is an observation which if omitted would have a considerable effect on the estimation process. An outlier is an observation at some 'distance' from the rest of the data and which appears to be inconsistent with the rest of the data. For both these types of observations it is necessary to decide what to do about the observation.

6.3   Homogeneous Poisson process (HPP)

An HPP is a renewal process for which the distribution of the times between events (failures) is assumed to be an exponential distribution, see Chapter 5, Subsection 5.4.2. This means that in any period the number of events (failures) will have a Poisson distribution. The model is relatively simple with the number of failures in non-overlapping intervals assumed to be independent and the probability of failure in any small interval of length dt is given by λdt, for some positive constant λ, which is the rate of occurrence of failures (ROCOF). Whilst for some electronic systems the exponential distribution is often a 'good' fit to the underlying lifetime distribution, for other types of components and systems it can be rather poor. Simple indices such as the *dispersion index* and the *coefficient of variation* can indicate rapidly the goodness-of-fit of the exponential distribution. The indices are based on the knowledge that if the distribution of the lifetimes is exponential then the coefficient of variation is 1, and if the distribution of the number of failures is Poisson then the index of dispersion is 1.

Definitions of indices
The following indices are defined for a random variable X with expectation $E(X)$ and variance $\text{Var}(X)$:

$$\text{Coefficient of variation} = \frac{\text{Var}(X)}{(E(X))^2};$$

$$\text{Index of dispersion} = \frac{\text{Var}(X)}{E(X)}.$$

If the index is greater than (less than) 1 the process is overdispersed (underdispersed) compared with the HPP. More formally tests can be devised for comparison of plausible alternatives to the HPP. Some of these will be discussed in Section 6.6. If there is a belief that the alternative is another distribution then, of course, one can compare fits. For example, if the alternative distribution is a Weibull distribution then comparison can be made using the likelihood functions since the Weibull distribution includes the exponential distribution as a special case (with the shape parameter equal to 1). Hence a hypothesis test can be performed to infer whether the shape parameter is 1 or not.

Example 6.3.1
Using the data on system A in Table 1.7 in Chapter 1, the times of failures (in hours) were

452   752   967   1256   1432   1999   2383   (3000).

The last time in parentheses was the end of the observation period. Times between failures (in hours) are therefore

452   300   215   289   176   567   384   617+.

The last time marked with + was a right censored observation. The sample mean is 428.57, the sample variance is 28945.60, and hence the estimate of the coefficient of variation is 0.16. Therefore the data are underdispersed compared with the exponential distribution. Hence one should consider other distributional forms or alternative processes to the HPP.

□

Example 6.3.2
Consider the data collected by Winfield (1988) on two systems, which are given in Table 6.1.

Analysing the data on the total number of failures per year gives the sample mean to be 3.96, the sample variance to be 6.68, and the index of dispersion is 1.69. Hence the model seems overdispersed compared with the Poisson distribution. Hence one should consider fitting other models with variable failure rates over time or other failure distributions.

□

Another method for assessing whether the lifetimes come from an exponential distribution is to plot minus the log of the reliability function of the lifetimes against the lifetimes. This should produce a linear graph, if the data come from an exponential distribution with expectation $1/\lambda$, and the slope of the graph is the hazard function, λ.

6.4   Non-parametric estimation of the ROCOF

A process is an HPP if the rate of occurrence of events (failures) does not change over time. Hence there is a need to be able to decide whether there has been a change in the rate of occurrence of failures (ROCOF or renewal density function) for the process. This can be done by graphical means by plotting an estimate of the ROCOF to assess whether there has been a change. The ROCOF can be estimated using two simple approaches, which are dependent on the form of the data, and are as follows:

(a) If the data are times of failure then the ROCOF can be estimated by the mean failure rate v_i, which is calculated by dividing the time axis into equal intervals (t_{i-1}, t_i) and is given by

$$v_i = \frac{\text{number of failures in } (t_{i-1}, t_i)}{t_i - t_{i-1}},$$

for interval i. If the estimate is obtained by using more than one copy of a system then the estimate must be adjusted by the number of systems 'at risk' of failure (and repair).

Table 6.1. Data collected on two systems

	Years since 1960									
	1	2	3	4	5	6	7	8	9	10
No. of failures										
Independent	2	3	7	3	1	2	5	1	1	0
Joint	0	2	0	0	2	1	0	3	2	1
Total	2	7	7	3	5	4	5	7	5	2
	11	12	13	14	15	16	17	18	19	20
No. of failures										
Independent	2	0	7	2	3	2	0	1	7	3
Joint	0	1	0	0	0	1	1	3	1	3
Total	2	2	7	2	3	4	2	7	9	9
	21	22	23	24	25	26				
No. of failures										
Independent	0	1	2	3	1	2				
Joint	0	0	0	0	0	0				
Total	0	1	2	3	1	2				

(b) If the data are the number of events in a given time interval the mean failure rate m_i for the ith interval is obtained by using a moving average. For the moving average of length k the mean failure rate is given by

$$m_i = \frac{n_{i-(k-1)/2} + n_{i+1-(k-1)/2} + \dots + n_i + \dots + n_{i+(k-1)/2}}{k},$$

when k is odd, and

$$m_i = \frac{n_{i-k/2} + 2n_{i+1-k/2} + \dots + 2n_i + \dots + n_{i+k/2}}{2k},$$

when k is even, where n_j is the number of events in the jth interval.

In both cases the general idea is to allow for the possibility of localized effects whilst removing the noise from the system. The intervals should

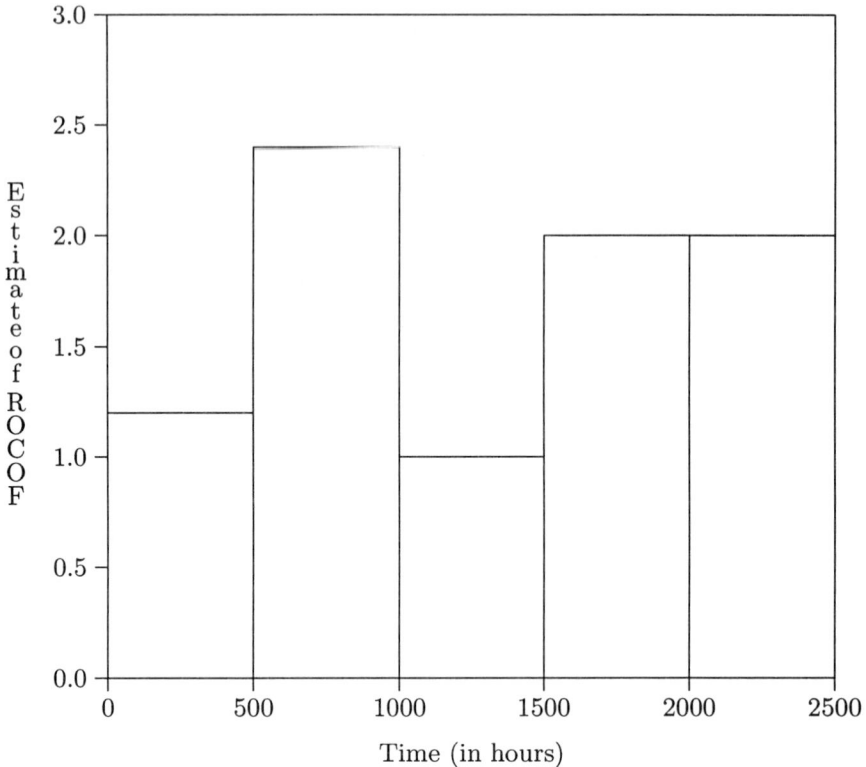

Fig. 6.1. Estimate of the ROCOF per 1000 hours for the repairable system failure data given in Table 1.7

therefore be chosen to be sufficiently long to give a reasonable estimate within the interval of the failure rate but not too long to dissipate local effects. In the case of the moving average k should take a low integer value.

Example 6.4.1
The plot of the mean failure rate per 1000 hours for the system failure data given in Table 1.7 in Chapter 1, Subsection 1.4.5, is given in Fig. 6.1 using intervals of 500 hours. This plot suggests that there is a failure rate of about 1.7 per 1000 hours.

□

Example 6.4.2
Using the data collected on the total failures for the two systems given in Table 6.1 a plot of the estimates of the ROCOF, the number of events occurring in a year, is given in Fig. 6.2 using the three- and five-point

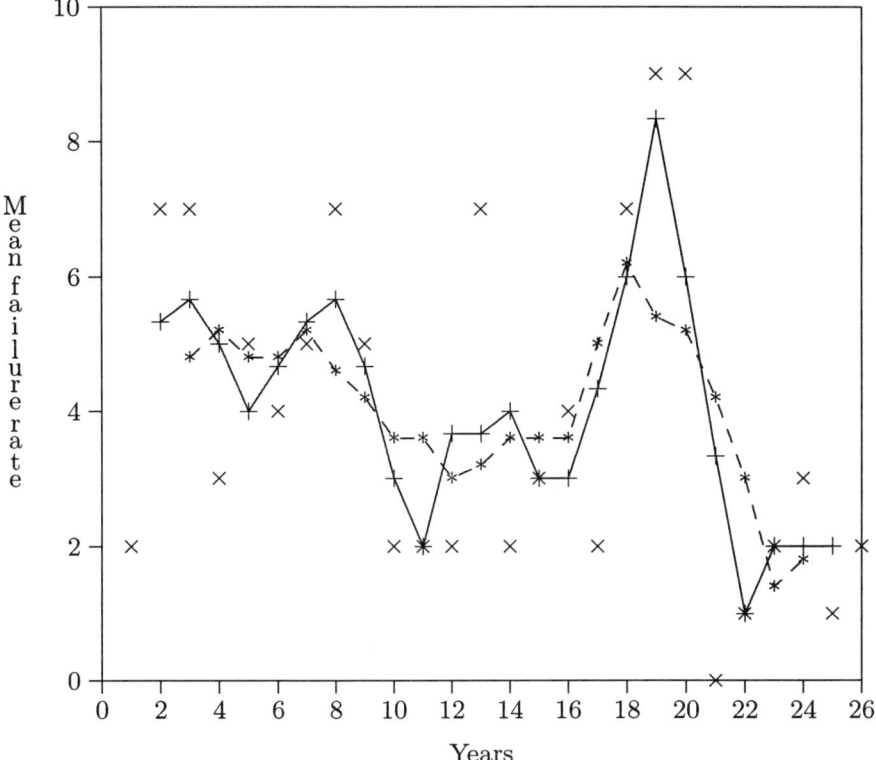

Fig. 6.2. Plot of the estimates of the ROCOF for the data collected on the total failures for the two systems given in Table 6.1 using three-(+) and five-(*) point moving averages

moving averages. The plot shows an initial rate of about 5 which falls to 2 except for two years (19 and 20) of poor performance. Hence it is obvious that the rate of occurrence is not constant throughout the 25 year period.

□

6.5 Non-parametric estimation of the renewal function

Nelson (1988) has suggested a simple non-parametric graphical method for estimating the expected cumulative number of failures $H(t)$ (the renewal function) for data from a number of copies of a repairable system. Recently he has proposed a method of calculating confidence intervals for this estimate.

Suppose a manufacturer of a system decided to assess the performance of the N copies of a system supplied to a customer and the following data are collected. When a failure of the system occurs in service the system is repaired and returned to service instantaneously. So there is a record for the nth system which consists of the failure times $t_{n,j}$ for the jth ordered failure. The record also contains the last time τ_n at which the nth system was withdrawn from service before the end of the period of data collection.

Nelson (1988) suggested estimating $H(t)$ by plotting the *mean cumulative number of failures* $C(t)$ at time t. This is calculated by dividing the time axis into N intervals by using the last times at which the systems were withdrawn from service before the end of the period of data collection. These are used after they have been ordered in decreasing magnitude so that

$$0 \leq \tau_N \leq \tau_{N-1} \leq \ldots \leq \tau_2 \leq \tau_1.$$

Then interval i is defined by $(\tau_{i+1}, \tau_i]$, with $\tau_{N+1} = 0$. Let $C_{i,n}$ be the number of failures in interval i for system n. Also let t be contained in the Ith interval. Then let $C_{I,n}(t)$ be the number of failures in the interval (τ_{I+1}, t) for system n.

With these definitions $C(t)$ is given by

$$
\begin{aligned}
C(t) \quad = \quad & [C_{N,N} + C_{N,N-1} + \ldots + C_{N,I} + \ldots + C_{N,1}]/N \\
& + [C_{N-1,N-1} + \ldots + C_{N-1,I} + \ldots + C_{N-1,1}]/(N-1) \\
& + [C_{N-2,N-2} + \ldots + C_{N-2,I} + \ldots + C_{N-2,1}]/(N-2) \\
& + \ldots \\
& + [C_{I,I}(t) + \ldots + C_{I,1}(t)]/I.
\end{aligned}
$$

Nelson also calculated the variance of this estimator, $\mathrm{Var}(C(t))$, as

$$
\begin{aligned}
\mathrm{Var}(C(t)) \quad = \quad & \sum_{i=I+1}^{N} \mathrm{Var}(C_i)/i + \mathrm{Var}(C_I(t))/I \\[6pt]
& + 2\sum_{i=I+1}^{N-1} \mathrm{Cov}(C_N, C_i)/N + 2\mathrm{Cov}(C_N, C_I(t))/N \\[6pt]
& + 2\sum_{i=I+1}^{N-2} \mathrm{Cov}(C_{N-1}, C_i)/(N-1) \\[6pt]
& + 2\mathrm{Cov}(C_{N-1}, C_I(t))/(N-1) \\
& + \ldots \\
& + 2\mathrm{Cov}(C_{I+1}, C_I(t))/(I+1),
\end{aligned}
$$

where $\mathrm{Var}(C_i)$ is the variance of $C_{i,n}$ and $\mathrm{Cov}(C_i, C_j)$ is the covariance of $C_{i,n}$ and $C_{j,n}$, for $i = j$. The variances $\mathrm{Var}(C_i)$ and the covariances $\mathrm{Cov}(C_i, C_j)$ are estimated by the usual sample estimates. Nelson suggested that this is a non-parametric equivalent to Greenwood's variance

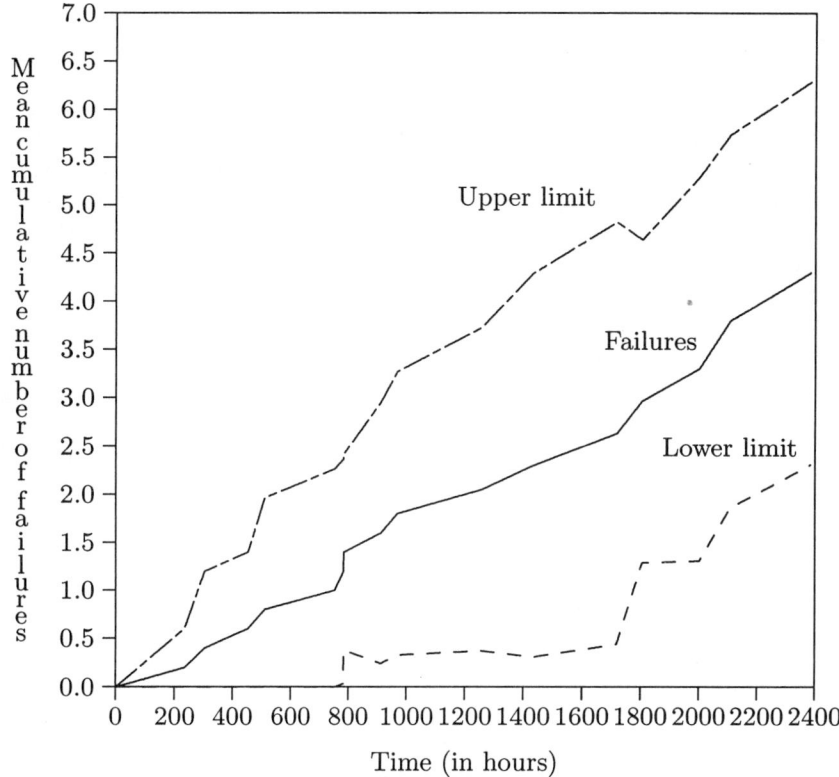

Fig. 6.3.   Mean cumulative number of failures for the repairable system failure data given in Table 1.7

for recurrence data. Limits can be calculated by adding and subtracting twice the standard error, the square root of the estimated value of the variance $\text{Var}(C(t))$.

Example 6.5.1
The plot of the mean cumulative number of failures for the system failure data given in Table 1.7 in Chapter 1, Subsection 1.4.5, is given in Fig. 6.3 with upper and lower limits. This plot suggests that there is a failure rate of about 1.8 per 1000 hours, which is consistent with the estimates of the ROCOF obtained in Example 6.4.1 and plotted in Fig. 6.1.

□

6.6   Trend tests

Confirmation of the visual inspection of graphical analysis can then be explored by testing for change. It is usually assumed in reliability analysis that the change will be monotonic and is either increasing or decreasing over time. Hence analysts will usually test for the trend. Obviously this is not the only possibility and one should be aware that the change may be non-monotonic. There are tests for both monotonic and non-monotonic trends. However, non-monotonic tests tend to be more specialized and hence only monotonic trends will be considered in this chapter.

There are a range of trend tests depending on the type of trend and process involved. No single test is uniformly powerful against all the possible range of alternatives. Hence the choice depends on what the analyst feels about both the process being studied and the alternative that could be considered. Two frequently used tests are the Laplace trend test, see Cox and Lewis (1966), and the MIL-HDBK-189 (1981) test. These are for comparing the HPP with the non-homogeneous Poisson process (NHPP).

Definition of the Laplace trend test
Suppose the data consist of times of events, so that t_i represents the time of the ith event and $(0, t_o)$ is the period of observation. Then the Laplace test statistic u_L is given by

$$u_L = \frac{\frac{\sum_{i=1}^{n} t_i}{n} - \frac{1}{2} t_o}{t_o \sqrt{\frac{1}{(12n)}}}.$$

If t_o is chosen so that observation stops at the time of the nth event then the above expression can be amended by replacing n by $n-1$ with $t_o = t_n$. Under the assumption that the process is an HPP the distribution of u_L rapidly approaches the standard normal distribution as the sample size increases so that large positive or negative values suggest the process is not an HPP. If the value is less than zero the process is improving, so that times between failures are increasing on average; if greater than zero the process is deteriorating, so that times between failures are decreasing on average. A hypothesis test can be performed using the result that u_L^2 has a χ^2 distribution with 1 degree of freedom, under the assumption of no trend for an HPP, and large values indicate departure from this assumption and are therefore consistent with the process being an NHPP.

Definition of the MIL-HDBK-189 trend test
Suppose the data consist of times of events, so that t_i represents the time of the ith event. Then the MIL-HDBK-189 test statistic u_M is given by

$$u_M = 2 \sum_{i=1}^{n-1} \log\left(\frac{t_n}{t_i}\right).$$

Now u_M, on the assumption that the process is an HPP, has a χ^2 distribution with $2(n-1)$ degrees of freedom. Again small and large values indicate departures from the assumption, with small values indicating deterioration and large values improvement, so a *two-sided* test is needed.

Both tests are powerful against a specific alternative hypothesis and will give limited protection for other alternatives. u_L is most powerful against the alternative of an NHPP with a ROCOF $\exp(a_0 + a_1 t)$, Cox and Lewis's model, whilst u_M is most powerful against the alternative of an NHPP with a ROCOF $\lambda \beta t^{\beta-1}$, Crow's model. As Stephens (1989) has pointed out the Laplace test may indicate a trend when there is none if the coefficient of variation is large, e.g. the Weibull and gamma distributions with shape parameters less than unity.

Example 6.6.1
Using the data on system A from Table 1.7 considered in Example 6.3.1 the Laplace test statistic is

$$u_L = -0.5495,$$

for the 7 failures and a period of observation 3000, and the MIL-HDBK-189 statistic is

$$u_M = 13.3098,$$

for the 7 failures. So the Laplace statistic indicates that there is some improvement in the system while the MIL-HDBK-189 statistic indicates that there is some deterioration.

Comparing u_L^2 with the χ^2 distribution with 1 degree of freedom there is no evidence of a trend. The u_M statistic should be compared with the χ^2 distribution with 14 degrees of freedom, and again there is no evidence of a trend. So both statistics are consistent with the null hypothesis that the model is an HPP.

$\square$

The tests are worth applying to chronological data as a first stage in any analysis. If the analyst has specific views about the alternatives then more appropriate tests might be considered. If for example the process is thought to be a renewal process rather than an HPP then the tests of Mann (1945) or Lewis and Robinson (1974) might be more appropriate.

The Mann test counts the number of reverse arrangements in the chronological order of the data. A reverse arrangement is whenever $X_i < X_j$ for $i < j$ in the list of data, $X_1, X_2, ..., X_n$. For the mean number of reversals, under the assumption of a renewal model with no trend, the expectation would be $n(n-1)/4$ with a variance equal to $(2n^3 + 3n^2 - 5n)/72$. For $n > 10$ the number of reversals under the assumption of a renewal process with no trend is approximately normal. For $n < 10$ Mann (1945) provides tables.

Example 6.6.2
Using the data on system A in Table 1.7 considered in Examples 6.3.1 and 6.6.1 the times between failure are 452, 300, 115, 289, 176, 567, 384, and 617. The number of reversals for each time is 2, 3, 5, 3, 3, 1, and 1, respectively. The total is 18. The expected number should be $8(8-1)/4 = 14$ and the variance is $(2 \times 8^3 + 3 \times 8^2 - 5 \times 8)/72 = 16.33$. It would appear again that there is no evidence of a trend in the data.

□

The Lewis and Robinson test adapts the Laplace test by dividing the test statistic by the square root of the coefficient of variation (CV) of the inter-arrival times to generalize from the HPP to a renewal process.

Example 6.6.3
Using, again, the data on system A in Table 1.7 considered in Examples 6.3.1, 6.6.1, and 6.6.2, the Lewis and Robinson test statistic would be

$$u_L/\sqrt{CV} = -0.5495/(\sqrt{0.1825}) = 1.286.$$

Again there is no apparent indication of a trend.

□

If a trend is established there are a number of approaches which may be taken. In Chapter 7 growth models are discussed, which are mainly NHPP models. It will be seen that these models assume that the change in the rate over time can be described by a specific model. There are alternative models which could be fitted, for example time series could be explored, see Walls and Bendell (1986).

6.7   Dependency

Another problem which calls into question the use of the HPP is dependence within the series. It is usual to assume that the dependency within a series is local. Hence if a sequence is $X_1, X_2, ..., X_k, ..., X_n$, then there may be dependency between the ith lifetime, X_i, and the $(i+r)$th lifetime, X_{i+r}, where r is a low integer value, usually $k = 1$ or 2.

A simple way of examining the relationship is to plot the data against the data after lagging. Obviously as the lag increases more points are lost and less significance can be attached to the plot. Dependency would be characterized by patterns in the data, though reliability data can be a problem because the distribution is often skewed. A suitable transformation of the data may produce plots which are easier to interpret. Natural choices for transformations are log, square root, or cube root, which may make the data more symmetric before plotting.

Whilst plotting is a first step it may be sensible to confirm the results of the plots and one approach is to use the auto-correlation function. The *auto-correlation function* of lag k, ρ_k, is defined as

$$\rho_k = \frac{\text{Cov}(X_i, X_{i+k})}{\sqrt{\text{Var}(X_i)\text{Var}(X_{i+k})}},$$

where $\text{Cov}(X_i, X_{i+k}) = E((X_i - E(X_i))(X_{i+k} - E(X_{i+k})))$ is the *covariance* of X_i and X_{i+k}. Cox and Lewis (1966) suggested using the estimator

$$\hat{\rho}_k = \frac{\sum_{i=1}^{n-k}(x_i - \bar{x}'_k)(x_{i+k} - \bar{x}''_k)}{\sqrt{\sum_{i=1}^{n-k}(x_i - \bar{x}'_k)^2 \sum_{i=1}^{n-k}(x_{i+k} - \bar{x}''_k)^2}},$$

where

$$\bar{x}'_k = \frac{\sum_{i=1}^{n-k} x_i}{n - k}$$

and

$$\bar{x}''_k = \frac{\sum_{i=1}^{n-k} x_{i+k}}{n - k}.$$

It should be noted that the auto-correlation function will only indicate whether there is a linear relationship between the serial lifetimes. The correlation coefficient is a measure of *linear* relationship, taking values ± 1 if there is a linear relationship between the two random variables. Hence if the relationship is non-linear this may not be noted by the use of the auto-correlation function. In such cases more complex models need to be considered which would be based on the analyst's knowledge. Another danger is that since the auto-correlation function is usually calculated for a series of lags it is possible that by chance one of the lags will give an auto-correlation which is significantly different from zero. Therefore great care has to be taken in interpretation.

Example 6.7.1
Using the data on system A in Table 1.7 considered in Examples 6.3.1 and 6.4.2, the lag 1 times are plotted against time and are given in Fig. 6.4. There does not appear to be any pattern in the plotted values, which is consistent with the auto-correlation estimate with a lag of 1 for the times to failure of the systems, which is 0.058. Also the auto-correlation with a lag of 2 is 0.35. Hence there does not seem to be any detectable dependence for these two lags.   □

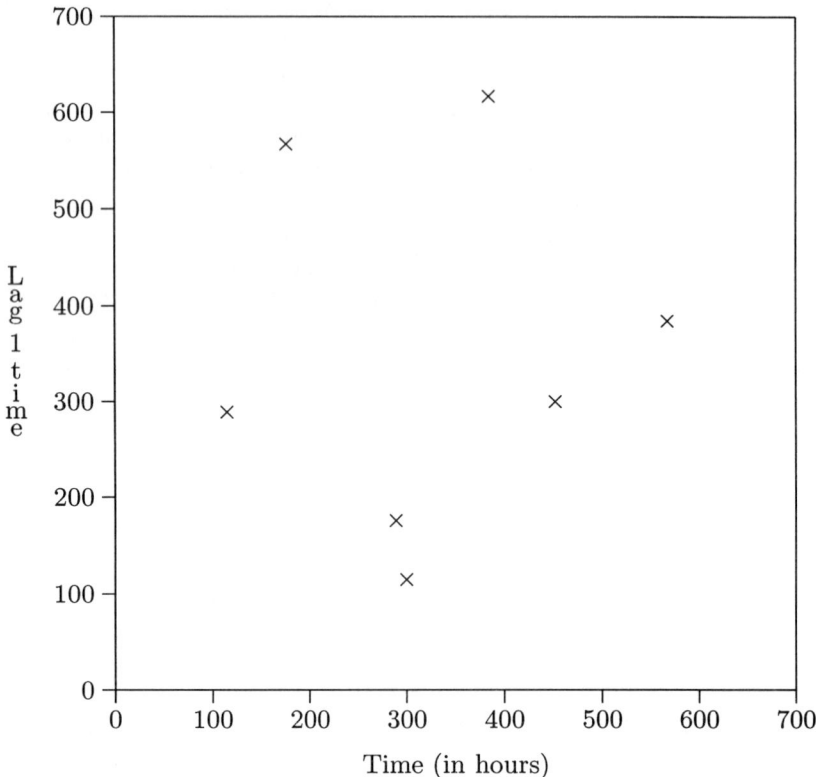

Fig. 6.4. Lag 1 plot for system A from the repairable system failure data given in Table 1.7

6.8   Stationarity

A process must be *stationary* to be an HPP. As Cox and Lewis (1966) observed, it is quite possible to define a stationary point process in a very restrictive way as 'the same invariance property must hold for the joint distribution of the number of events in a set of k fixed intervals, for all $k = 1, 2, ...$'. However, such a definition will be difficult to verify with empirical data. Often a weaker definition of stationarity is taken in which only the first two properties of the process are invariant under a time shift. It should also be noted that the time at which the series is first observed may affect whether the data appear stationary or not. For example, if the observer starts observing the process at the time a renewal occurs then the process will not be stationary. For it is unlikely that there will be a failure within a short epoch of the start. Cox and Lewis (1966) pursued the concept of stationarity of a process through the application of Fourier transforms. These transforms can be very helpful in understanding a process and its properties. This is outside the scope of the current text and readers are referred to Cox and Lewis (1966) for more details.

6.9   Fitting other models

So far in this chapter the question of whether it would be appropriate to fit an HPP to the data has been examined. If this model is rejected as inappropriate then an alternative will have to be considered. The reason for rejection will often suggest a possible alternative. Though an alternative to the parametric approach of considering an alternative parametric process model and estimating the parameters of the model is to make no specific model assumptions and to use the non-parametric estimation approaches in Sections 6.4 and 6.5. Before proceeding, however, a warning should be given that there may be insufficient data to reject the HPP. So there is a danger of underfitting, that is fitting a simpler model than is required to the data.

If it is established that the times between failures are non-exponential then one could contemplate two alternatives. Firstly a more appropriate distribution may be fitted to the data such as the Weibull distribution. If this is the case then one could use one of the parametric estimation methods suggested in Chapter 2 to fit the distribution. As an alternative to fitting a distribution one might contemplate fitting a more complex stochastic process. This may be more difficult. For example, suppose one decides that a superimposition of renewal processes (SRP) was an appropriate model; then one would need to decide about the number of processes involved and the lifetime distribution for each of the processes, though one might assume the processes are identical. Strong assumptions may be required and it may not be possible to justify them. For deciding on the choice of the appropriate model see the comments in Section 6.2.

If the departure from an HPP is due to varying rates of occurrence of failures then one might consider using the NIIPP or apply the growth models described in Chapter 7. The estimating of the parameters of Crow's and Cox and Lewis's models, described in Chapter 5, Example 5.4.2, are described by Crowder et al. (1991). Also Crow's model is described in Chapter 7, Subsection 7.2.4. If the main concern is dependency the models discussed in Chapter 8 would be worth considering.

Another model which may be appropriate, if there is more than one possible cause or mode of failure, is the competing risk model. It is assumed that the causes compete in that the first cause to occur causes failure to the component. Failure having occurred, the other causes are censored. So if there are m causes, $i = 1, 2, ..., m$, for the jth component then the lifetime of the component is the minimum of the lifetimes to failure for each of the causes, $t_{j_1}, t_{j_2}, ..., t_{j_m}$, so the component's lifetime, t_j, is given by

$$t_j = \min_{1 \leq i \leq m} (t_{j_i}).$$

The simplest analysis of competing risks is to treat each failure as a failure event for the specific cause involved and as a censored observation for the other causes. Hence the likelihood function would be composed of probability density functions relating to failures from specific causes and reliability functions for the censored data. In the above notation, assuming it is the ith cause which leads to failure, then the contribution to the likelihood function for the jth component is

$$f_{ij}(t_j) \prod_{l=j} R_{lj}(t_j),$$

where $f_{ij}(t_j)$ is the probability density function of the ith cause for the jth component and $R_{lj}(t_j)$ is the reliability function for the lth cause for the jth component. The estimation methods described in Chapter 2 can be used to obtain estimates of the parameters. Obviously there is a need to specify the distributions for all the causes. For further details of competing risk models see Cox and Oakes (1984) or Gail (1975).

6.10   Analysing complex systems

Previously in this chapter the interest has centred on the components' behaviour. In the remaining sections of this chapter the problem of analysing a whole system is considered. The analysis of the system will depend on the nature of the data available. Usually the data are limited either to the system as a whole or alternatively to the components which compose it. It is a rare occasion that the data are available on both component behaviour and system behaviour.

If the data are only available at the system level then there is limited point to treating them at a more sophisticated level than as a component. Hence the analysis should proceed as already described. There may be occasions when it is possible to provide insight to alternative models for the system based on the system behaviour.

If data are only available at the component level then the approach is based on combining this information in an appropriate fashion. For example, if the behaviour of a series system with instantaneous repair is considered then the times between failure are an SRP. The model for each process can be derived from the component data. The performance of the whole system might be assessed through its renewal density, h_s, which would be the addition of the component renewal densities, h_j, $j = 1, 2, ..., m$, so that h_s is given by

$$h_s = \sum_{i=1}^{m} h_j.$$

An example of its use is given in the case study in Chapter 10, Section 10.3.

In the case when information is available both at the system and component level then the analyst is in a fortunate position. It is then possible to validate the model. One could obtain the expected performance of the system through the model of the system and the components' behaviour and compare this with the observed system performance.

6.11 Markovian models

The performance of many large systems in reliability is described by Markovian models. Several reasons are put forward for the selection of such models. The prime reason is the simplicity of the models. Another consideration is the quality of the data to support the model. If it is regarded as poor, then it may be felt that more elaborate models cannot be validated. For example, the data available may be limited to mean lifetimes of components and mean repair times. Frequently there will be limited information on the systems' performance or structure. The information may be hypothetical or engineering judgement about the system based on FTA, see Chapter 4, Section 4.8.

Care has to be taken in the selection of the model and in some cases it might be appropriate to explore a range of models which may be plausible rather than a single model. This may form a sensitivity analysis. If information is available at both system and component level then it may be possible to assess the likelihood of various structures for the system. This might be achieved through comparing observed performance against expected performance of the system given the components' behaviour and the structure for the system. This will be relatively rare.

If the data are not available on the system's performance then the estimates will have to be made on the behaviour of the components and an assumed structure. Estimation will therefore depend on the available information at the component level. If only point estimates are available for the expected lifetimes of components then it does seem that the assumption of an exponential distribution may be a possible approach. Hence one may use the approaches described in Chapter 5 to find the system's performance in both the transitory and steady state. If the data at the component level are available in more detail then it may be possible to explore the component's distribution. The analysis should follow the method described earlier in this chapter. If the distribution can be accepted as exponential then one can proceed as described above. If the data are not from an exponential distribution then it may not be possible to obtain exact distributional forms for the system's performance without substantial numerical calculation. Two approaches which could be employed are the use of simulation or the construction of moments of the system's performance. Simulation carries with it the twofold drawback of the computational time to obtain estimates and then the reliability of the estimates.

6.12　Discussion

This chapter has considered the analysis of repairable systems. The assumption has been made that repair is instantaneous. This may at first seem to be a restrictive (unrealistic) assumption. However, this is not the case in many practical studies as either the repair times are short compared with the times between failures or only the operating time of the system is of interest. This second situation is realistic when there is no significant degradation in the system when it is not operating.

　　Initially it was assumed that the data came from the simplest model, the HPP. This is a reasonable model for a system over the 'early' part of the life of a system. The appropriateness of this model may be tested first by the use of indices and then using tests for a trend. In Section 6.6 two trend tests were presented and compared. These tests are powerful against different specific alternative hypotheses. The problem of dependency in a series of times between events was considered in Section 6.7 using the auto-correlation function. The auto-correlation function is only a measure of linear relationship, so if the relationship is non-linear it may not be detected by the use of the auto-correlation function. However, one should be aware that there is the danger of underfitting if there is insufficient evidence to reject the HPP.

　　The behaviour of the system can be investigated by estimating the ROCOF and the renewal function. Non-parametric estimators for both these functions were presented.

　　The chapter then progressed to consider more complex data. Choosing the most appropriate model will depend on the objective of the study. Often it will require judgements about the plausibility of certain models which may be solely based on engineering knowledge.

7
Growth and adaptive models

7.1 Introduction

The aim of most reliability studies is the improvement of the performance of the system under study. Hence it is reasonable to believe that over a period of time there should or ought to be an improvement in the lifetime of components/systems. Growth models were originally introduced to describe such improvement. The early models were constructed on the basis of expected improvement. It should be appreciated that the 'improvements' are not always successful and in some cases the system's performance can be downgraded by the 'improvements'. Hence the optimism of the growth models should be countered by the pragmatic realization of the possible degradation of a system's performance over time.

The title growth model in reliability has become attached to a set of models which allow for changes in the lifetimes of components over time. These models will include trend models as a special case. In fact trend models are monotonic growth models. Most models assume that growth is monotonic and usually that it is positive with respect to the variable of interest. (The time between failure is increasing, the rate of events decreasing, etc.)

It is possible to divide the growth models into two types: those designed for hardware systems and those for software systems. The chapter will discuss hardware models such as Duane's and Crow's and software models such as those of Jelinski and Moranda and Littlewood and Verrall. However, the models are not restricted to the area for which they were designed but are often used interchangeably to assess or predict growth for hardware and software. This chapter will, however, not explore software reliability since this is an extensive field of study in itself; for details of the subject see Smith (1985) or Musa *et al.* (1987).

Both sets of growth models might be described in terms of their *expected lifetime*, $E(T)$, the *expected (cumulative) number of failures*, $E(N(t))$ (introduced in Chapter 5, Subsection 5.4.1, as the renewal function), or the *cumulative mean time between failures*, $CMTBF(t)$, at a time t, say. For example, a growth model might be described as a model for which the expected lifetime $E(T)$, expressed as a function of overall time t, is given by

$$E(T) = g(t, \beta),$$

where g is a function of t and β, a set of parameters of the function, which may be known or unknown.

Often the data will be aggregated from either field trials or laboratory experiments with only the number of failures in given time intervals recorded. Sometimes the data will be taken from a single series of components. Obviously the choice of model will depend on the data available. Hence the nature of available data will be explored and the models relevant to such data will be considered.

7.2   Growth models

In this section the models will be investigated. This will include a description of the model and how estimates of the parameters are obtained. The applicability of the model will also be explored.

7.2.1   DUANE'S MODEL

Based on the concept of learning curves Duane (1964) suggested that the *cumulative mean time between failure*, $CMTBF(t) = t/E(N(t))$, can be modelled as

$$CMTBF(t) = \frac{t}{E(N(t))} = Kt^{\alpha},$$

so that $E(N(t))$, the expected number of failures at time t, is given by

$$E(N(t)) = \frac{1}{K}t^{1-\alpha},$$

where K and α are parameters and t represents the time on test. K is sometimes referred to as the location parameter and α as the growth parameter. Most analysts concentrate on α since this represents the improvement exhibited by the system/component.

Estimation for this model is usually performed by graphical means since if $\log{(\widehat{CMTBF}(t))} = \log{(t/N(t))}$, where $N(t)$ is the *cumulative number of failures* at time t, is plotted against $\log t$ and provided the model is appropriate then the graph should be approximately a straight line. The slope of the graph will be α and the intercept will be $\log K$. This plot is known as a *Duane* plot. Obviously it is possible to use formal estimation methods, such as least squares or weighted least squares.

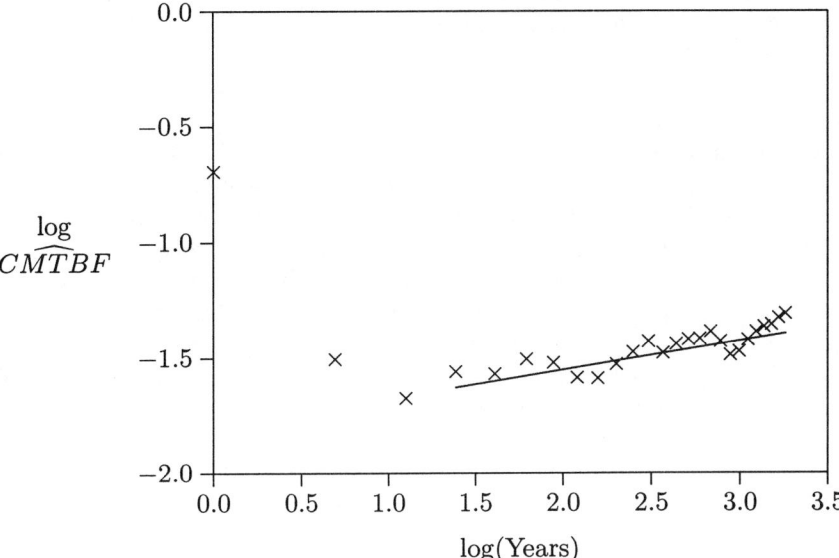

Fig. 7.1. Plot of log($C\widehat{MBT}F$) against log time with the 'best' line fitted by 'eye' using Duane's model with the data collected on the total failures for the two systems given in Table 6.1

Example 7.2.1
Using the data collected on the total failures for the two systems given in Table 6.1 of Chapter 6, the Duane plot of log $(C\widehat{MTB}F(\text{years}))$ against log(years) is given in Fig. 7.1. The parameters of Duane's model have been fitted by 'eye' with the following results:

$$\hat{\alpha} = 0.125 \text{ and } \log \hat{K} = -1.8,$$

which indicate a 'small' growth in reliability.

□

It is the graphical aspect of this model which has probably made it so popular, especially the relatively easy estimation. It has been used by many analysts. However, the model does suffer from a number of disadvantages. It assumes that the change in performance is monotonically smooth. Technological changes and procedural changes tend to occur at distinct points of time, hence producing a series of jumps in systems performance. It may be well approximated by a smooth function but there will be discrepancies. More worrying, though, is the assumption of monotonicity, since it is frequently assumed that α is positive, which implies that the system is always improving. Whilst many technological changes may be positive in their effect there is always the possibility that some of the effect of changes will be negative. Hence it would seem unlikely to see monotonic change throughout the length of a project. Again it is plausible to argue that the average change will hopefully be fairly smooth.

As well as these problems various authors (Crow, 1974; Sheppard, 1983) have commented on the estimation problems. Firstly there is a need to define t, the time on test. When does recording of the time start: at inception of the project, at the design stage, or at some other time? In Example 7.2.1 it does appear that the effect of the low number of failures in the first year has a sizeable effect on the estimation. Also what should be the basis for time measurement: calendar time or time on test or some other scale? Having clarified these points there are several questions about the reliability of the estimates. Several problems arise associated with the early points on the plot which may have a very significant effect on estimates. Also, given the nature of the estimates it is not possible to give a measure of precision to the estimates, unless more formal methods are used. Hence it is not possible to assess the reliance which should be placed on them.

The technique can be applied to both aggregate data or single series data, though it must be admitted that the latter will be very unreliable given the precision that can be attached to estimating $CMTBF(t)$ on a single series.

7.2.2   IBM MODEL

One of the problems associated with Duane's model is that at $t = 0$ the $CMTBF(t)$ is also zero. This does not generally appeal to many analysts since it implies an infinite failure rate. The IBM model (Rosner, 1961) overcomes this problem. It is assumed that there are two types of failure: inherent failures occurring at a constant rate and design or manufacturing defects. These latter defects are removed through time and so the failure rate decreases. The expected cumulative number of failures, $E(N(t))$, is given by

$$E(N(t)) = At + B\left(1 - \exp(-Ct)\right),$$

where A, B, and C are parameters. A represents the inherent failure rate whereas B and C both refer to the defect rate. As $t \to 0$ the second term disappears. Hence over time the effect of the defects disappears and the inherent failure rate is dominant. Estimation of A, B, and C can be by least squares estimation. It must be noted, though, that the function is non-linear and hence requires an iterative approach, and there may be problems with the slow convergence of the estimation procedure.

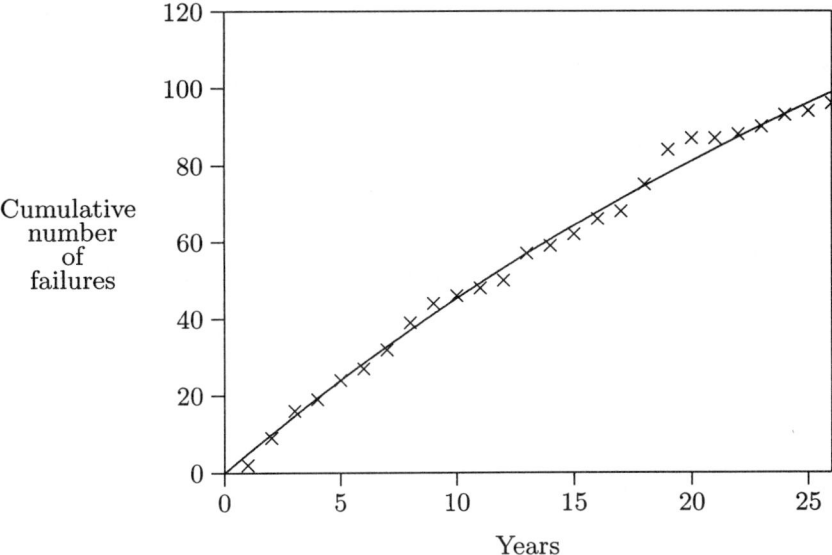

Fig. 7.2. Plot of the cumulative number of failures against time (in years) with the 'best' fitted equation for the IBM model with the data collected on the total failures for the two systems given in Table 6.1

Example 7.2.2
Fitting the model to the data collected on the total failures for the two systems given in Table 6.1 resulted in slow convergence and after a large number of iterations the values obtained were $A = 0.262$, $B = 189.97$, and $C = 0.0255$. Fig. 7.2 gives the plot of $N(t)$, the cumulative number of failures at time t, with the IBM model with a curve drawn using the parameters above. The fit appears reasonable, though at the upper end there is an indication of a change of rate in the process which the model cannot reflect. According to the interpretation of the model it does appear that after 26 years some of the initial defects still require to be removed. Hence there is some concern at least about the interpretation of the data.

□

Again the model assumes a smooth monotonic change in the rate. With C assumed to be positive the rate is decreasing and the system's reliability is improving. Hence the previous comments apply, though unlike Duane's model the choice of t is not so crucial since B can be regarded as a scaling factor. So it is more reliable than Duane's model and with the method of estimation usually employed it is possible to construct standard errors.

Therefore it may be regarded as a better model to fit if the data were consistent with the supposed monotonic form.

As with Duane's model it can be applied to both aggregate and single series data in its present form.

7.2.3 RATE MODELS

There are a number of models which are based on the growth rate of a specific attribute. Examples are Lloyd and Lipow's (1962), Aroef's, and the exponential rate models. Again these models assume monotonically smooth changes in the rate, usually assumed to be decreasing with time. This is usually expressed as

$$\frac{dY(t)}{dt} = k(Y(t), t, \beta),$$

where $Y(t)$ is a suitable measure of growth at time t, usually $CMTBF(t)$, and k is some decreasing function of t whose parameters are β. Each model has been applied at least once to a set of selected data. Little can be said in their favour except there are obviously some very specific applications for which they are appropriate. Estimation again is often by least squares though other approaches could be considered.

In more detail the models are as follows.

Lloyd and Lipow's rate model

$$\frac{dCMTBF(t)}{dt} = \frac{B}{t^2}.$$

Hence

$$CMTBF(t) = \begin{cases} A - \dfrac{B}{t}, & \text{for } t > B/A, \\ 0, & \text{for } t < B/A, \end{cases}$$

and estimation can be carried out by least squares using the transformation of t to $1/t$.

Aroef's rate model

$$\frac{dCMTBF(t)}{dt} = \frac{B \cdot CMTBF(t)}{t^2}.$$

Hence

$$CMTBF(t) = A \exp(-B/t).$$

Taking logs yields

$$\log(CMTBF(t)) = \log A - B/t$$

and again parameters can be estimated using least squares with $\log(CMTBF(t))$ and using the transformation of t to $1/t$.

Exponential rate model

$$\frac{dCMTBF(t)}{dt} = B.CMTBF(t).$$

Hence

$$CMTBF(t) = A\exp(Bt).$$

Taking logs yields

$$\log(CMTBF(t)) = \log A + Bt$$

and again parameters can be estimated using least squares.

Example 7.2.3

Using the data collected on the total failures for the two systems given in Table 6.1, the estimates of the parameters are

Lloyd and Lipow's: $\hat{A} = 0.2144$; $\hat{B} = -0.1889$.

Aroef's: $\log \hat{A} = -1.5081$; $\hat{B} = -0.4959$.

Exponential: $\log \hat{A} = -1.4716$; $\hat{B} = 0.0027$.

Fig. 7.3 shows a plot of the $\widehat{CMTBF}(t)$ of the data with the curves from the three models superimposed. Again none of the models is capable of reflecting the changes in the $CMTBF(t)$ of the data. The best fit of the three is Aroef's model, though this is still relatively poor.

□

7.2.4 CROW'S MODEL

Previous models have solely considered the path of a characteristic of the underlying process, either the CMTBF or the cumulative rate. They have not been concerned with the underlying process. However, if a distribution is not assumed in reliability the implication is that it is exponential. This would imply a Poisson process, yet this process would have a variable rate and therefore would be a non-homogeneous Poisson process (NHPP), see Chapter 5, Subsection 5.4.5. Crow (1974) recognized this and suggested that an NHPP with a renewal density function (ROCOF) given by

$$h(t) = \lambda\beta t^{\beta-1}, \text{ where } \lambda \text{ and } \beta > 0,$$

was the equivalent of Duane's model, see Chapter 5, Subsection 5.4.5. The maximum likelihood estimates for λ and β were derived by Crow (1974) for both lifetime and aggregate data.

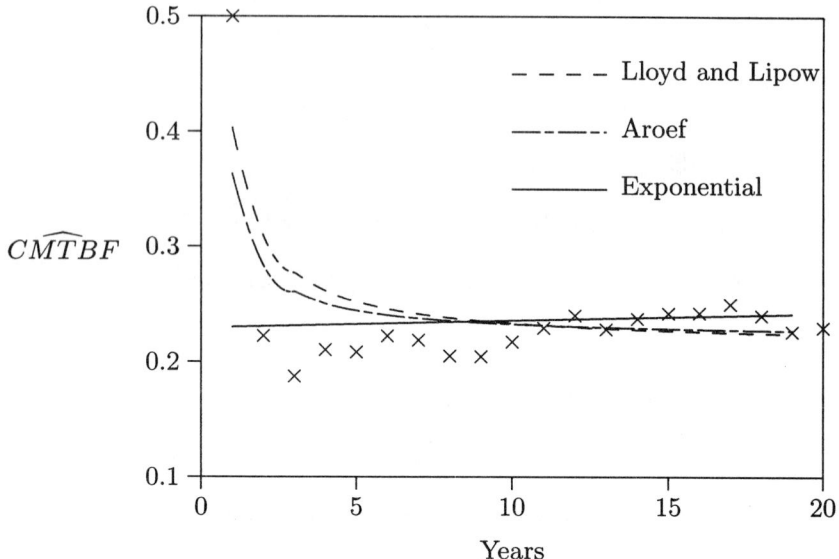

Fig. 7.3. Plot of the estimate of the cumulative mean time between failures against time (in years) with the 'best' fitted equations for Lloyd and Lipow's, Aroef's, and the exponential rate models with the data collected on the total failures for the two systems given in Table 6.1

For lifetime data when observations are made from time 0 to T with failures at $t_i, i = 1, ..., n$, the estimator $\hat{\lambda}$ is given by

$$\hat{\lambda} = \frac{n}{T^{\hat{\beta}}},$$

and the estimator $\hat{\beta}$ is given by

$$\hat{\beta} = \frac{n}{\sum_{i=1}^{n} \log(T/t_i)}.$$

For aggregate data when only the number of failures is known in given time intervals then it is possible to use the result which states that for an NHPP the number of events in a time interval (t_i, t_{i+1}) has a Poisson distribution with expectation $\lambda(t_{i+1}^{\beta} - t_i^{\beta})$, see Cox and Lewis (1966). Suppose there are p intervals which end at $t_i, i = 1, ..., p$, each containing k_i failures. Then the equations for the maximum likelihood estimators are

$$\hat{\lambda} = \frac{n}{t_p^{\hat{\beta}}}$$

and

$$k_1 \log t_1 + \sum_{i=2}^{p} k_i \left(\frac{t_i^{\hat{\beta}} \log t_i - t_{i-1}^{\hat{\beta}} \log t_{i-1}}{t_i^{\hat{\beta}} - t_{i-1}^{\hat{\beta}}} \right) = n \log t_p,$$

where $n = \sum_{i=1}^{p} k_i$.

Example 7.2.4
Using the data collected on the total failures for the two systems given in Table 6.1 with equal time periods of unit length, the second maximum likelihood equation becomes

$$\sum_{i=2}^{p} k_i \left(\frac{i^{\hat{\beta}} \log (i) - (i-1)^{\hat{\beta}} \log (i-1)}{i^{\hat{\beta}} - (i-1)^{\hat{\beta}}} \right) = n \log p.$$

Hence $\hat{\lambda} = 5.64$ and $\hat{\beta} = 0.87$.

□

Obviously given the NHPP model it is possible to consider Bayesian estimators for the parameters, see Basu and Rigda (1986). In such cases there is a need to specify a prior distribution for the parameters.

As formulated this model again suffers from the problems previously expressed. However, Duane's model is only being used as an example. The rate need not be monotonic and it is possible to suggest alternative formulations for the rate.

A number of authors have considered fitting such models. For example, Cox and Lewis (1966) discuss in detail the model with a ROCOF given by $\exp(a_0 + a_1 t)$ when considering fitting a trend. This model has been generalized by Lewis (1972) to include a quadratic term and by Lewis and Shedler (1976) for further polynomial terms. Lewis (1972) showed that fitting the quadratic term could have a dramatic effect on the estimate of the a_1. Lee (1980) considered fitting a composite of Crow's and Cox and Lewis's models resulting in a ROCOF of $\beta t^{\beta-1} \exp (a_0 + a_1 t)$. Cozzolino (1968) produced an 'initial defects' model which was equivalent to Cox and Lewis's model which was then generalized.

A more interesting approach was taken by Braun and Paine (1977), who considered a model in which the rate was based on the number of failures uncovered by time t. This model, though, appeared to perform badly in practice.

It is also possible to suggest adaptive models in which the rate discontinuously changes as in a step function. There are, however, generally problems with the selection of change points. One solution is that the failure times define the change points.

The use of the NHPP for growth models is only one of many possible stochastic process models which could be employed, see Chapter 5. It is therefore possible to consider other stochastic models.

7.2.5 SOFTWARE MODELS

The IBM model can be regarded as both a hardware and software model. The failure rate is dependent on an inherent failure rate and a decreasing failure rate dependent on the number of original defects. It is the second part which many subsequent authors have used to describe software failure models based on 'bugs' (or defects). However, unlike the IBM model they have not assumed a smooth monotonic decrease in the rate. Usually they have assumed that each discovery of a failure has caused the rate to be modified. It is also accepted in some of the models that the rate can increase as well as decrease.

All the software reliability growth models subsequently described assume that the data come from a single series of data. This is not as restrictive as might initially appear since what inherently is being assumed is that many versions of the software are being run but once a fault is discovered all versions are corrected. Hence the models are being applied to aggregate data more generally. These extensions will not be considered.

7.2.6 JELINSKI AND MORANDA'S MODEL

This model, proposed by Jelinski and Moranda (1972), was the most common software reliability model used during the 1980s. The model assumed that there are N defects. Each defect is assumed to be equally likely to cause the system to fail. The times at which the failures occur are assumed to be independent. Once a defect causes a failure it is immediately corrected and the number of defects decreases by one. The failure rate is supposed to be proportional to the number of defects still remaining.

The failure rate at the beginning, if there are N defects, can be expressed as

$$u_1 = N\phi,$$

where N is the number of defects and ϕ is a constant of proportionality. After the ith defect at time t_i is discovered the rate will be

$$u_i = (N - i)\phi.$$

A graphical representation of this is given in Fig. 7.4, for $N = 23$ and $\phi = 1/2.3$. It is again assumed that the times between failures have an exponential distribution. Hence the probability density function of the distribution of x_i, the time between the $(i - 1)$th failure at time t_{i-1} and the ith failure at time t_i, is given by

$$f(x_i) = (N - i + 1)\phi \exp\left(-(N - i + 1)\phi x_i\right), \text{ for } x_i > 0.$$

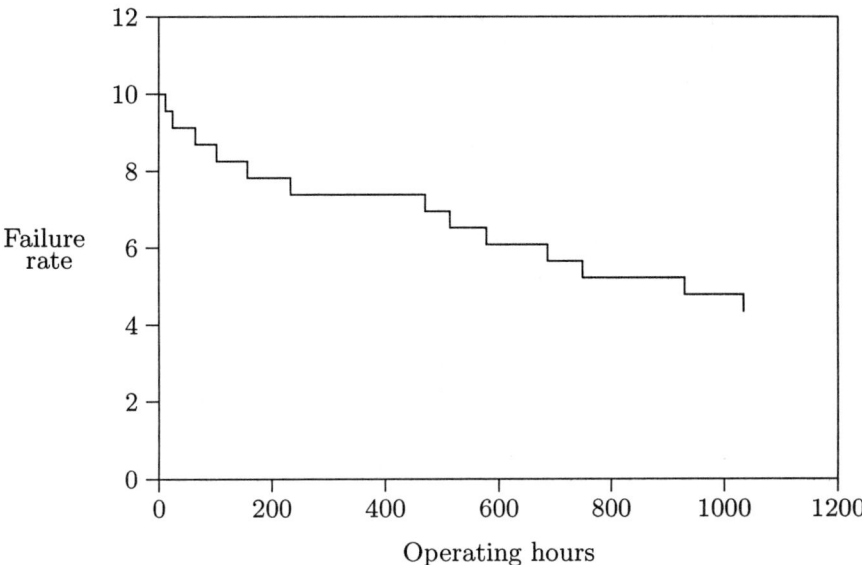

Fig. 7.4. Example of the failure rate (u_i) for Jelinski and Moranda's model for 23 defects and constant $\phi = 1/2.3$.

The distribution is therefore defined by two parameters N and ϕ, since it is unlikely that it will be known how many defects there are initially.

The maximum likelihood estimator of N is the value which satisfies

$$n \log \left(1 - \frac{t_n}{s_N} \right) + \log \left(\frac{N}{N-n} \right) > 0$$

and

$$n \log \left(1 + \frac{t_n}{s_N} \right) + \log \left(\frac{N-n+1}{N+1} \right) > 0,$$

where n is the observed number of failures, t_n is the time at which the last failure was observed, and s_N is given by

$$s_N = (N-n)t_n + \sum_{i=1}^{n} t_i.$$

The estimator of ϕ is then obtained as

$$\hat{\phi} = \frac{n}{(\hat{N} - n)t_n + \sum_{i=1}^{n} t_i}.$$

The model can be criticized on many grounds. Firstly the assumption of monotonic improvement is felt by some analysts to be inappropriate.

Software correction, as with technological improvement in hardware, is equally likely to downgrade performance by the introduction of new faults. Littlewood (1981) suggested that since some faults are likely to be more frequently encountered than others, this therefore will lead to a different rate of fault occurrence. Others feel that the constant rate of failure between failures arising from the assumption of an exponential distribution is unrealistic.

Forman and Singpurwalla (1977) found that the maximum likelihood estimator of N was unstable and, if the number of failures is high, can be misleading. This may be interpreted as a shortcoming of maximum likelihood estimators.

A refinement of this model is due to Musa (1975, 1980), who introduced a number of additional factors to the model. These included an 'error reduction factor' B and a 'testing compression factor' C_t. The effect of C_t on the mean time to failure (MTTF) can be described as

$$MTTF = \frac{T_0 \exp(C_t)}{N_0 T_0},$$

where T_0 is the initial mean time to failure of the system and N_0 is the initial number of faults. The model, though refined, has not tackled some of the major drawbacks of Jelinski and Moranda's model and especially some problems with estimation.

7.2.7   LITTLEWOOD AND VERRALL'S MODEL

A slightly different approach is taken by Littlewood and Verrall (1973) in that they consider not the number of faults within the system but the inputs to the system which may cause system failure. The system is assumed to be subject to a sequence of inputs some of which will cause the system to fail in its current state. The inputs are assumed to be random and it is therefore assumed that the probability density function of the distribution of x_i, the time between the $(i-1)$th failure and ith failure, is given by

$$f(x_i) = \mu_i \exp(-\mu_i x_i), \text{ for } x_i > 0,$$

where μ_i is the failure rate between the $(i-1)$th and ith failure. Once a failure occurs the original fault is removed and the underlying failure rate μ_i is changed.

Littlewood and Verrall (1973) assumed that the μ_i were realizations from a gamma distribution with a scale parameter given by a parametric function, $\phi(i)$, and shape parameter A. If it assumed that $\phi(i) = (B_1+B_2i)$, then the probability density function of the distribution of μ_i is given by

$$f(\mu_i) = \frac{(B_1 + B_2i)^A \mu_i^{A-1} \exp(-(B_1 + B_2i)\mu_i)}{\Gamma(A)}, \text{ for } \mu_i > 0.$$

This choice of distribution assumes that stochastically the failure rate decreases provided B_2 is positive. It does, however, mean that at some failure points the rate can increase, though on average it is decreasing. The distribution of X_i, the time between the $(i-1)$th and ith failure, can be obtained directly by integrating out μ_i using Bayes' theorem and is found to have a probability density function given by

$$f(x_i) = \frac{A(B_1 + B_2i)^A}{(x_i + B_1 + B_2i)^{A+1}}, \text{ for } x_i > 0.$$

Littlewood and Verrall (1973) proposed a standard Bayesian estimation procedure to find parameter values. However, they presented an equivalent procedure which provides point estimates. The method requires the calculation of u_i and $Y(u)$. The u_i can be obtained using

$$u_i = 1 - \frac{\log\left(\prod_{m=1}^{i-1} k_m\right)}{\log\left(\prod_{m=1}^{i} k_m\right)},$$

where $k_m = \phi(m)/(x_m + \phi(m))$. The $Y(u)$ are the proportion of the x_i which satisfy $x_i < x_i(u)$, where $x_i(u)$ is a $100u\%$ confidence bound defined by

$$P(X_i < x_i(u)) = u.$$

The u_i should have a uniform distribution on $(0, 1)$ if the model is appropriate. The $Y(u)$ should also have a uniform distribution on $(0, 1)$. A plot of u_i against $Y(u)$ should be a straight line with slope 1. The closeness of the plot to the line indicates the quality of the fit for the model. The measure of the closeness of the fit could be a 'goodness-of-fit' statistic such as the Kolmogorov–Smirnov statistic, which is the maximum vertical distance between the plot and the slope. Littlewood and Verrall (1973) suggested that selection of estimates can be achieved through minimizing an appropriate measure of 'goodness-of-fit', and the Kolmogorov–Smirnov statistic is convenient.

An interesting aspect of this model is that the failure rate $\mu(t)$ at time t, after n failures, is estimated by

$$\hat{\mu}(t) = \frac{n+1}{[\hat{B}_1 + \hat{B}_2(n+1)]} \cdot \frac{1}{\prod_{i=1}^{n+1}\left(\log\left(\hat{B}_1 + \hat{B}_2i + x_i\right) - \log\left(\hat{B}_1 + \hat{B}_2i\right)\right)},$$

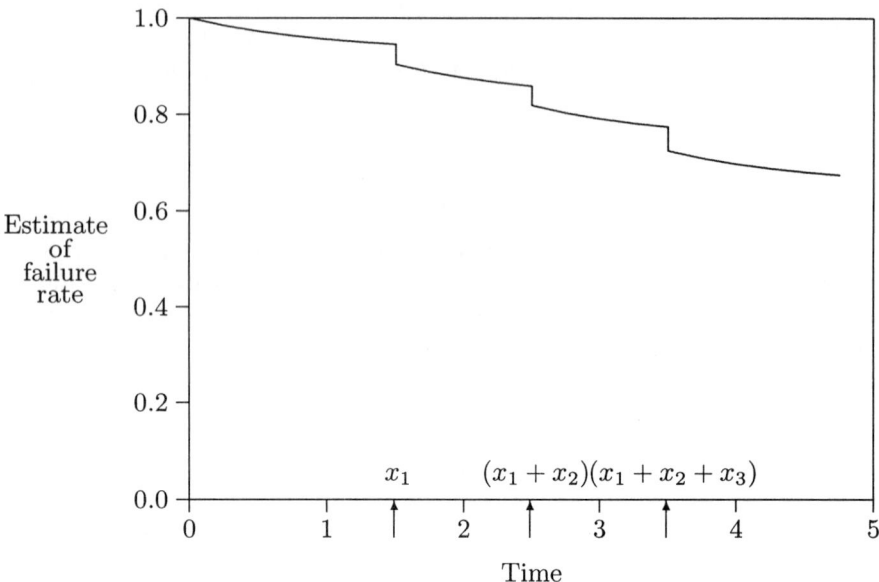

Fig. 7.5.   Example of an estimate of the failure rate for Littlewood and Verrall's model

where

$$x_{n+1} = t - \sum_{i=1}^{n} x_i$$

is the time since the last failure. A graphical representation of this is given in Fig. 7.5 and it should be noted that the failure rate between failures is not constant but decreases. This is due to the increasing confidence which comes with a failure free period. Criticism of this model has been more limited since it overcomes some of the disadvantages of previous models. However, some analysts believe that the model should allow for perfect 'debugging' by assuming a quantum of probability for zero failure rate.

7.2.8   LITTLEWOOD'S MODEL

An alternative model based on a superimposition of renewal processes (SRP) was suggested by Littlewood (1981) to get round the problem of assuming that all faults have the same propensity to cause failure. Littlewood supposed that if the failure rate for defect j was ϕ_j, then the overall rate for failure, μ_i, after i faults have been detected and corrected, would be

$$\mu_i = \sum_{j=1}^{N-i} \phi_j,$$

if the underlying distributions were assumed to be exponential. Again it was assumed that ϕ_j were realizations from a gamma distribution with shape parameter A and scale parameter B.

Hence the distribution of x_i, the time between the $(i-1)$th and ith failure, has a probability density function given by

$$f(x_i) = \frac{(N-i+1)A(B+t')^{(N-i+1)A}}{(B+t'+x_i)^{(N-i+1)A+1}}, \text{ for } x_i > 0,$$

where t' is the time to detect and correct the first $i-1$ faults. The model does assume that failures do not introduce new faults. Here there is a strict monotonic decrease in the estimated future rate. There are other ways of extending the Littlewood and Verrall model which will not be explored.

7.3 Other models

In this chapter the two main types of growth model, which commonly occur in reliability, have been considered. The first set assumed that the failure rate, or another characteristic, depends on a function of the time t. For most of these models an implicit assumption was that the underlying distribution was exponential. The second set of models assumed that a change in the failure rate occurred upon failure. Again the underlying distribution was assumed to be exponential.

These two types of models are not the only types which can be considered when attempting to study changes in the failure rate of components. A number of other approaches could be considered. Since the aim is to model the growth of failure rate the main impetus must be to discover the pattern or structure in the data. There are many statistical techniques which could assist in this process of discovery. Simple exploratory techniques such as plotting could be used. It may be informative to plot the $CMTBF$ or ROCOF to see if there is a pattern, see Chapter 6. The pattern could then be smoothed or an appropriate model fitted to the data. Alternatively one could apply time series analysis to the data, an approach which has been attractive to those studying software reliability such as Walls and Bendell (1986) and Soyer (1986). It is also possible to extend the models presented. For the first set other functional forms could be considered, and for both sets it would seem sensible to explore distributions other than the exponential.

7.4   Availability and applicability of data

Generally data have been available on hardware failures to develop the functional models. The success therefore for such models has been well established. Obviously the literature only rarely reports the cases where the model is inappropriate. Percra (1990) studies a range of models and comes to the conclusion that Duane's and Crow's models fit well the data he considered. Other authors, though, have reported that the underpinning assumptions of Duane's model are inappropriate for data for hardware reliability. Halliday and Devereux (1983) indicated that the 'region of non-conformity' to Duane's model was extensive. Much of the data had to be discounted to fit the model, which is clearly not a good idea.

For the software reliability models the story is less positive. Initially there was a dearth of data with commercial sensitivity being the major reason for non-release. Subsequently the databases were set up to collect data under a number of programmes including the UK Alvey programme. Data analysis has been less rewarding with many models found to be wanting. At the present time, whilst there is still interest in software reliability modelling, there does not appear to be any indications of success in this endeavour.

7.5   Discussion

Throughout this chapter it has been stressed that the models tend to assume growth. For some models, especially Jelinski and Moranda's model, if the data do not show growth then the estimation procedure is a problem and may lead to infinite values. However, this is not the only cause for concern. Jewell (1978a,b) has shown that estimators of $CMTBF(t)$ will tend to increase in t even if the underlying process is stationary. This was established by both analytic and simulation methods for the HPP. Hence Jewell suggested that there is a need to consider time horizons of 5 or 10 times the MTTF to obtain a reasonable estimate of the parameters. Obviously the more parameters that are fitted the more care has to be taken to ensure that the period of data collection is sufficiently long. This chapter has concentrated on describing the various models used. The models are generally used in reliability growth management with the objective of deciding on the time to release the product. Hence usually the models are fitted and extrapolation made from them. So the main aim is not to fit the current data but to estimate the time at which the reliability will meet a specific target, but it appears that generally such estimates will be poor, see Leach (1988).

8
Dependency analysis

8.1 Introduction

So far it has been assumed that the lifetimes of components (or subsystems) are independent. This chapter explores approaches which do not make this assumption. A recent review of the area has been carried out by Ansell and Walls (1990, 1992). Modelling and the analysis of data from systems where components (subsystems) are dependent are becoming increasingly important as it is realized that dependency may significantly affect the reliability of the system. This has been particularly noted for safety systems when redundancy did not produce the expected level of performance owing to dependency amongst the components.

The assumption of dependency leads to a number of problems including its definition. In statistical terms two events are independent if their joint probability is the product of their marginal probabilities so that for two events A and B

$$P(AB) = P(A)P(B).$$

Hence dependent events are those for which

$$P(AB) \neq P(A)P(B).$$

This may not be regarded as particularly helpful in comprehending when events are dependent or alternatively in attempting to model dependent events.

In engineering applications dependency is usually recognized by causal linkage such as mechanical wear, electrical stress, or a common environment. A typical example is the most frequent type of model used in reliability engineering for dependency, the common cause model, where failures are linked back to a single 'root' event. If the mode of failure is the same for the components then this is sometimes termed common mode failure. (There is some disagreement over the definitions of these two models.)

The previous comments on the quality of data, see Chapter 1, Section 1.5, are especially important in the context of studies of dependency. The development of appropriate models has often been hindered by the lack of data. Most of the data available for studies of dependency come from secondary sources such as inventory records, maintenance, and incident reports.

Obviously the identification of dependency is the first step in the analysis. Many engineers have written on identification from a qualitative standpoint, examining incident reports to discover the nature and form of dependencies, see Amendola (1988) and Ballard (1989). This has tended to focus on coincidental failures and hence the important common cause model. In this chapter the emphasis is on quantitative techniques for identification. These may lead to a more general view of dependency, allowing for other forms beyond coincidental failure models. It is important, though, that forms of dependency uncovered by quantitative methods are supported through a qualitative causality.

The models which are usually applied in reliability studies of dependency are the 'cut-off' model, the beta factor model and its extensions, and the state of knowledge models. These models concentrate on common cause failure. Underpinning these models is the assumption that joint failures can be classified by analysts as independent or dependent events. This subjective classification can bring into doubt the validity of the models. This and other aspects of each of these models will be discussed.

There are a number of models developed by statisticians for modelling, identifying, and assessing dependency. These tend to make simplifying assumptions about the context. Whilst such models are more mathematically sophisticated than those currently used it is doubtful whether the data collected conform to the assumptions made. The types of model to be considered can be broadly described as renewal, environmental, and multivariate, though there is some overlap in using these definitions.

8.2   The nature of dependency

It is not possible to give a definition beyond the simple statistical definition given in the introduction in Section 8.1, which will cover all possible forms of dependency which may be encountered in reliability. It is, though, worth considering the broad range of possible descriptions of dependency.

In this chapter the main concern is with repairable systems with at least two components (subsystems) which may be subject to joint failures. At this stage interest is not focused on performance at the system level but on the components (subsystems). This may be referred to as between systems dependency. Alternatively for a sequence of components there may be dependency within the sequence, with x_i, the lifetime of component i, dependent on x_j, the lifetime of component j, for some j, where $i > j$.

Table 8.1. A characterization of dependency

Causal link	Nature of link	Effect on item	System process	Status of resulting failure	Description
Environmental	Deterministic	Instantaneous	Dynamic	Revealed	Common cause
Structural	Stochastic	Delayed	Static	Unrevealed	Common mode
Component					Time dependent
Manufacture					
Procedural					
Design					

Table 8.1 gives a possible characterization of dependency within reliability. A number of possible causal links exist, such as environment, structure, manufacture, procedure, etc. Environment is taken to cover a wide range of possible connections including the physical context, mechanical wear, and outside effects which might even include procedural aspects of maintenance. The form of linkage between the processes may be deterministic, in that certain types of events will cause the components to fail, or stochastic in that failure of one component may increase the chance of the other component failing. The latter case implies the possibility of dependency through time. A special case is the cascade model where a component may cause subsequent failure of other components. However, dependency through time can assume a greater generality. Hence included in time dependency would be the undesired side effect of maintenance in which repairs, or replacements, may affect the subsequent performance of components sharing the same context. It may be that failures are recognized as they occur or are noticed at the next inspection. Thus models for dependency should allow for instantaneous or delayed failures. It is relatively simple to adapt the models to consider unrevealed failures. Fig. 8.1 illustrates the types of models under consideration. Note that joint failure does not imply dependent failure. Crude data recording frequently means that care has to be taken as to whether failures are dependent or not. In case (a) are the failures at t_A and t_B dependent? This equally applies to time dependency. It may be that some failures do cause subsequent failures in other components but this is not always true. In case (b) only the failures at t_{a_1} and t_{b_1} cause failures in the other process. Great care must always be taken in examining data. Engineering judgement is paramount but statistical techniques may be helpful in identifying possible dependencies. Few authors within reliability have considered this important latter problem.

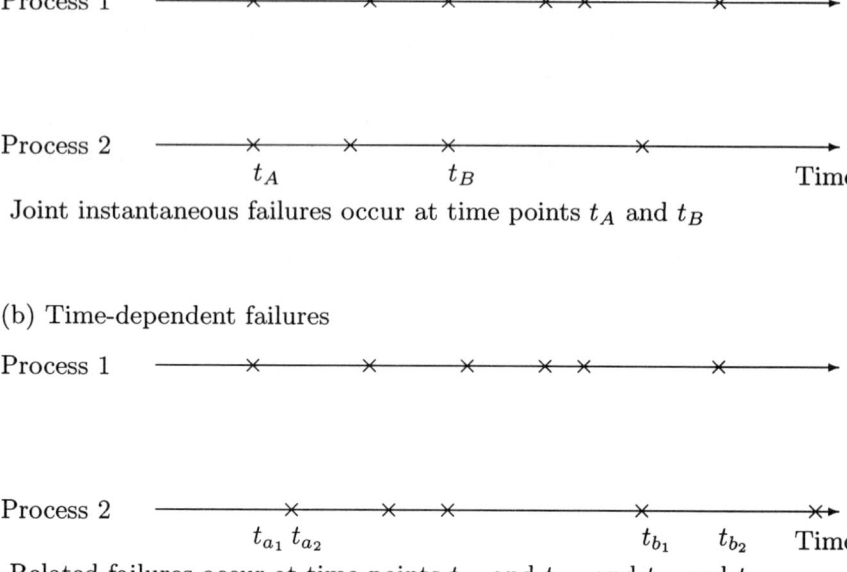

(a) Joint instantaneous failures

Process 1

Process 2

t_A t_B Time

Joint instantaneous failures occur at time points t_A and t_B

(b) Time-dependent failures

Process 1

Process 2

t_{a_1} t_{a_2} t_{b_1} t_{b_2} Time

Related failures occur at time points t_{a_1} and t_{a_2}, and t_{b_1} and t_{b_2}

Fig. 8.1. Two models for repairable systems with dependency. × denotes an
event or failure

8.3   Data

In Chapter 1 the problems associated with the collection of reliability data
were described. These comments are particularly pertinent in the context
of dependency studies. The secondary nature of most of the data does
cause problems for the analyst.

In dependency studies qualitative data are useful to gain understanding
about the nature and form of dependency. In the nuclear industry this
can be gleaned for example from incident reports. Amendola (1988) and
others have written about the use of qualitative data from incident and
alert reports in analysis. They are especially helpful in recognizing and
understanding the underlying failure mechanism. Quantitative data to
some extent may also be used to assist identification as well as giving
insight into the frequency of the effect.

Incident reports do not provide good indicators of the frequency of
dependent failures since they provide information about a special event, an
alert, or incident. The alternative is the maintenance report which provides

some indication of frequency, though often the data are aggregated. This can be a major drawback since most statistical approaches produced are based on the availability of data on lifetimes.

8.4 Identification of dependency

Identification of dependency is obviously an important aspect of any study. A number of papers outline engineering techniques, for example Amendola (1988), Crellin *et al.* (1988), Fleming *et al.* (1983), Games *et al.* (1985), and Humphreys *et al.* (1987). It is also possible to investigate dependency using simple statistical techniques such as scatter plots and correlation.

For a series of events plots of the times between failures against lagged time between failures may reveal patterns due to dependency. Serial correlation may also reveal certain forms of dependency and tests exist for auto-correlation, see Chapter 6, Subsection 6.7. Between series similar approaches can be used. However, there may be difficulties in deciding which values from the series to use in calculation of the correlation coefficients; see later comments in Subsection 8.5.5 on multivariate models for lifetime distributions. There are a number of other graphical methods for the identification of dependency, see Cox and Lewis (1966).

Generally dependency is highlighted through a change in the rate of failure. In common cause failure coincident failures are an obvious example. Similarly in other models the effect is to increase or even decrease the occurrence rate from that expected.

A possible way therefore to identify failure is to assume no dependency and see if there are departures from the model. An example is the proposal by Walls and Bendell (1989). They describe a graphical technique which can be used to explore field data. The data are assumed to be aggregate data, collected either on a daily basis or within some other time window. These daily failure times for systems of identical or related components are observations from a non-homogeneous Poisson process (NHPP) model with a ROCOF $h(t) = \lambda\beta t^{\beta-1}$, see Example 5.4.2 in Chapter 5. This is based on the assumption that after repair the systems reliability remains essentially unchanged. The properties of this NHPP model are discussed in Bendell and Walls (1985) who fit this model to the data under the null hypothesis of no dependency. The data are examined for significant departures from the null model by constructing confidence intervals about the estimated expected number of failures, obtained through simulation. Particularly high failure rates compared with the null model would lead to points falling outside the confidence interval and so indicate departure which may be due to dependency.

Crow's estimators, see Chapter 7, Subsection 7.2.4, can be used to obtain estimates of the parameter values. The confidence intervals may be obtained as explained in Walls and Bendell (1989), though for simplicity one can approximate by using the normal approximation to the Poisson distribution. This will be a reasonable guide provided the Poisson rate is greater than 9. If the rate is less than 9 for any interval then it is possible to find the exact confidence interval. This leads to the approximate $100(1 - \alpha)\%$ confidence interval

$$\hat{\lambda}t^{\hat{\beta}} \pm z_{\alpha/2}\sqrt{\hat{\lambda}t^{\hat{\beta}}},$$

where $z_{\alpha/2}$ is the $100(1-\alpha/2)\%$ centile of the standard normal distribution.

Example 8.4.1
To illustrate this the data collected on the total failures for the two systems given in Table 6.1 of Chapter 6 are used. The joint failures are treated as two failures in Walls and Bendell's approach. Crow's estimates of the parameter values from Chapter 7 are $\hat{\beta} = 0.87$ and $\hat{\lambda} = 5.64$. Fig. 8.2 gives the plot of the cumulative number of failures assuming an NHPP model and the 95% confidence intervals, both the simulated, as suggested by Walls and Bendell, and the approximate ones given above. Note that the data according to this approach do not exhibit any effect of dependency except possibly for the first observation.

□

Walls and Bendell's approach was developed as an exploratory tool. Departures from the null model should be explored with the possibility that there might be dependency amongst components. The dependency would only be accepted if qualitative justification could also be found. The approach was applied to a number of component data sets and did succeed in highlighting certain cases of dependency, see Walls (1994). Subsequent simulation work on the approach against plausible alternatives has been disappointing. The sample size has to be reasonably large to ensure detection of dependency. Cosgrove (1992), however, has shown that shock models with Weibull times between shocks may indicate dependency for values of the shape parameters less than one. Hence care should always be taken in interpretation.

An alternative approach is to fit a model in which dependency is represented by a parameter or more than one parameter. It would then be possible to test to see if the parameter was significantly different from zero and hence there would be indications of dependency of the specified form. A number of possible models which allow this type of analysis are now discussed.

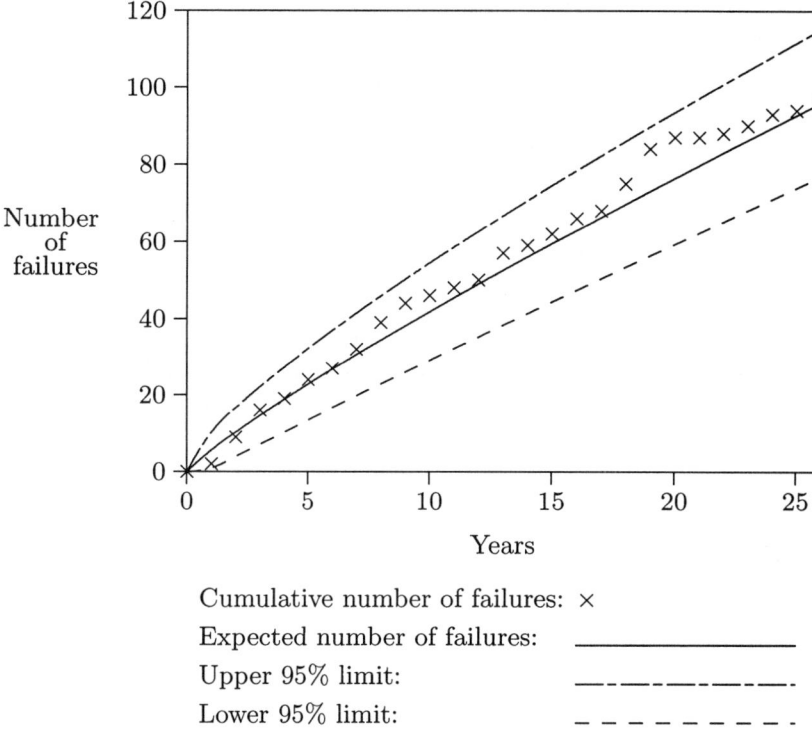

Number of failures

Years

Cumulative number of failures: ×
Expected number of failures:
Upper 95% limit:
Lower 95% limit:

Fig. 8.2. Cumulative number of failures for aggregate data

8.5 Models for dependency

8.5.1 'CUT-OFF' METHOD

A serious misconception has arisen in the use of the guide by Bourne *et al.* (1981). The survey information contained in this handbook was intended to be used by designers and operators to gain an indication of the levels of reliability which might be achieved by various system designs, such as simple redundant systems, partially diverse systems, or two separate diverse systems. The precise ranges of values for the different types of system configurations are given in Bourne *et al.* (1981). These values are based on historical data and engineering judgement, and hence it is generally accepted that the low failure probabilities for large complex systems are speculative.

Some reliability engineers have unfortunately tended to use the figures as if the choice of a specific design ensures the specified level of performance. Worse still, others have spent time on arguing that their design fits a particular class of design in order to claim the specified level of reliability. It is important that the figures are used as a guide to designers and that a full analysis of a system is carried out before reliability figures are claimed for it.

8.5.2   BETA FACTOR MODEL

The original beta factor model, see Fleming (1975), assumed that two identical pumps were taken to be operating in the same environment and hence subject to similar conditions. It was assumed that the probability of failure of a pump, p, could be partitioned into the probability of independent failure, p_I, and the probability of dependent failure, p_C, i.e.

$$p = p_I + p_C.$$

Therefore the probability of failure of two identical components would be $p_I^2 + p_C$. Further the probability p_C could be expressed as βp, which is the reason for the name of the model.

There is no problem with the formulation. However, to use this model one must be capable of defining the partition. Since it is subjective, different analyses can result in different values of p_C and p_I. Thus it runs counter to the usual statistical notation which does not suffer from this subjectiveness. In the statistical model there is no partition but a joint probability of failure is used which is denoted as p_U. From above p_U could be expressed as

$$p_U = p_I^2 + p_C.$$

If the two identical components are statistically independent then $p_U = p_I^2$ and $p_C = 0$. It is possible to derive p_I and p_C from observation of p and p_U and thus avoid this subjectiveness.

The beta factor model has been extended to processes. It is assumed that the rates can be partitioned in the same way as probabilities without taking account of the underlying distributions. Hence the rate of failure for a component, μ, is assumed to be decomposed into μ_I, the independent failure rate, and μ_C, the dependent failure rate, so that

$$\mu = \mu_I + \mu_C.$$

Again the estimation requires a subjective partition of the data to estimate μ_I and μ_C. It should be noted that a higher reliability for the system will be achieved if more joint failures are deemed independent rather than dependent.

The rates, μ_I and μ_C, are estimated by using

$$\hat{\mu}_I = \frac{N_I(t)}{2t} \ ,$$

where $N_I(t)$ is the number of independent failures in the time period till t and $2t$ represents the total time on test for the two components, and

$$\hat{\mu}_C = \frac{N_C(t)}{t} \ ,$$

where $N_C(t)$ is the number of joint failures in the time period till t and t represents the time on test for the two components jointly.

Example 8.5.1
Using the data collected on failures for the two systems given in Table 6.1 the estimates of the rates are as follows:

$$\hat{\mu}_I = 61/(26 \times 2) = 1.173$$

and

$$\hat{\mu}_C = 13/26 = 0.5,$$

and the estimates of the sample variances are 4.315 and 0.66, respectively. Hence the estimates of the indices of dispersion are 1.83 and 1.32 for independent and joint failures, respectively.

□

In most cases when this approach is used specific distributions for the underlying failure process are not assumed. This makes it hard to assess the estimators and therefore they can be regarded as rather a crude method. This will be discussed further in the section on multivariate models.

A variety of possible interpretations of the model are plausible. The usual simple assumption of an exponential distribution for the times between failures for the underlying processes can unfortunately lead to a non-standard analysis being necessary. Bueno and Sherwin (1990) illustrate that different assumptions about the model can lead to significantly different estimates being obtained.

If the underlying distribution is not exponential then two problems arise. The exponential distribution's 'lack of memory' property will be lost and hence the process will have to be modelled in greater detail. Also truncation will become a problem. For example, if joint failures take place then it is a reasonable assumption that the independent failures have been truncated. Hence estimation beyond the exponential will become more complex.

In Example 8.5.1 the index of dispersion indicates that the data may be overdispersed compared with the Poisson distribution, which suggests some evidence of a non-exponential distribution. In practice, though, the assumption of an exponential distribution for the failure times is often regarded as plausible. For electronic components the failure times are commonly found to be (approximately) exponentially distributed.

8.5.3   EXTENSIONS TO THE BETA FACTOR MODEL

It is possible to apply the beta factor model to systems with higher levels of redundancy, from two-component systems to n-component systems. However, in such cases it provides a conservative model since it assumes that the common cause will make all components in the group fail.

There are a number of extensions of the beta factor model. These include the multiple parameter model, also known as the multiple Greek letter model, and the partial beta factor model. The multiple parameter model introduces further parameters to distinguish between common cause events affecting different numbers of components in highly redundant systems. As an example, the total failure of a single item in a 2-out-of-3 system, p, can be decomposed so that

$$p = p_1 + p_2 + p_3,$$

where p_1 is the single independent failure probability, p_2 is the double dependent failure probability, and p_3 is the triple dependent failure probability. Equivalently

$$p = (1 - \beta)p + \beta(1 - \tau)p + \beta\tau p,$$

where

$$\beta = \frac{p_2 + p_3}{p_1 + p_2 + p_3}$$

and

$$\tau = \frac{p_3}{p_2 + p_3}.$$

There are alternative possible formulations and again the approach has been expanded to deal with processes. The method of fitting parameters is identical to the case of two components.

Example 8.5.2

This example is based on data given in Hughes (1988): 500 systems consisting of three identical components are tested and their performance is given in Table 8.2.

Table 8.2. Performance of 500 three-component systems

System events	Number of components failing	Component events Successes	Failures
484	0	1452	0
13	1	26	13
2	2	2	4
1	3	0	3

From these data the estimates are

$$\hat{p}_1 = (13 \times 1)/(3 \times 500) = 8.67 \times 10^{-3}$$
$$\hat{p}_2 = (2 \times 2)/(3 \times 500) = 2.67 \times 10^{-3}$$

and

$$\hat{p}_3 = (1 \times 3)/(3 \times 500) = 2 \times 10^{-3}$$

or

$$\hat{\beta} = (2 \times 2 + 3)/(13 + 2 \times 2 + 3) = 7/20 = 0.35$$

and

$$\hat{\tau} = 3/(2 \times 2 + 3) = 3/7 = 0.43.$$

$\square$

The partial beta factor model allows for the incorporation of engineering judgement in a structured manner. The failure probability of a system is decomposed according to the built-in defences for the potential common cause dependency based on engineering knowledge. Hence the beta factor β is split into separate beta factors $\beta_1, \beta_2, \ldots, \beta_m$. The model may be additive or multiplicative as indicated below:

$$\text{Additive model: } \beta = \sum_{i=1}^{m} \beta_i;$$

$$\text{Multiplicative model: } \beta = \bigcap_{i=1}^{m} \beta_i.$$

The β_i will be either estimated from data or based on engineering judgement. Edwards (1988, 1989) described in detail the form of the analysis including the data requirements. The partial beta factor model has been adapted by the introduction of trigger mechanisms, see Games and Ballard (1988). However, the simple model represents the extent of sophistication likely to be encountered within UK reliability analyses.

8.5.4 STATE OF KNOWLEDGE MODELS

The state of knowledge models cover a more diverse set of models. These generally assume that a range of environments exist for a component owing to either conditions or usage. Each environment has a specific reliability. A typical example is the distributed failure probability model, see Hughes (1988). Suppose that an environment is denoted by w; then the probability of failure in that environment will depend on w and is denoted by $q(w)$. If the probability of being in that environment w is $p(w)$ then the probability of failure of the component p is

$$p = \sum_{w} q(w)p(w),$$

which is obtained by summing over all environments. If several identical components exist in the same environment then it is assumed that they are independent and have the same probability of failure. Hence dependency is defined solely in terms of the environment. If there are n components in the environment w their joint probability of failure will be $(q(w))^n$. The joint probability of failure, p_j, for the n components will be

$$p_j = \sum_{w} q(w)^n p(w).$$

The model can be easily extended to a continuous environment, see Hughes (1988). Hughes' model can be regarded as a generalization of Vesely's binomial failure rate model, see Vesely (1977) and Atwood (1986).

If the environments are known then estimation is simply restricted to monitoring behaviour in the environment. If the environments are not known then problems may arise with identification. Bayesian approaches are generally used, see Apostolakis and Moineni (1987).

Example 8.5.3
The data in Example 8.5.2 are used with the added assumption that the there are environments A and B. Environment A is defined so that only 0 or 1 failures are possible and environment B is defined so that only 2 or 3 failures are possible. Then the estimates of the probabilities are given by

$$\hat{P}(A) = 497/500$$

and
$$\hat{P}(B) = 3/500,$$

so that

$$\hat{P}(\text{component failure} \mid A) = 13/3 \times 497 = 8.73 \times 10^{-3}$$

and
$$\hat{P}(\text{component failure} \mid B) = 7/9 = 7.78 \times 10^{-1}.$$

Obviously such an analysis is very crude as Hughes (1988) comments and could be refined by a better definition of the environments independent of the data and possibly a fuller Bayesian analysis as presented in Hughes (1989).

□

Defining the 'environments' is critical to such analysis. This again allows for subjective judgement to enter a study since the data have to be disaggregated into these 'environments' and the separate probabilities calculated for each. The main criticism of such models must be that the level of complexity increases rapidly without necessarily yielding greater insight into the form of dependency. Yet if the 'environments' are based on sound judgements then they ought to be incorporated explicitly within the model instead of introducing extra layers without further clarification of the failure mechanism.

There are strong similarities between the state of knowledge model and the partial beta factor model. Environments can be taken to be equivalent to the defence mechanisms employed in the engineer's judgement. So the same criticism can be levelled against state of knowledge models. Again little attention is often paid to the underlying distributions.

8.5.5   MULTIVARIATE MODELS FOR LIFETIME DISTRIBUTIONS

Multivariate models are widely used by statisticians to model dependency and technically all the models discussed can be described under this umbrella. However, in this group specific distributions for lifetimes will be discussed. Initially the section will be concerned with the joint distribution of the first time to failure of two or more components.

Shaked (1982) suggested the following ratio as a measure of dependency

$$r(t) = \frac{R(t)}{\sum_{j=1}^{p} R_j(t_j)},$$

where $R(t)$ is the joint reliability function defined as $R(t) = P(T > t)$, where $T > t$ implies $T_j > t_j$ for all j, and $R_j(t_j)$ is the reliability function for the jth component. Obviously if $r(t) = 1$ then the components are independent. This can be used as a general description, but it is necessary to specify the marginal distributions for practical purposes. Generally it is assumed in the multivariate models that the failure time for each component is known.

As might be expected the early work in this area started with bivariate models. A number of authors suggested possible bivariate exponential distributions, see Downton (1970), Freund (1961), Gumbel (1960) and Marshall and Olkin (1967). These distributions were generally constructed on the assumption that the individual components would have exponential marginal distributions. A number of the models generalized the concept of 'lack of memory' property of the univariate exponential distribution, e.g. the bivariate exponential distribution of Marshall and Olkin (1967) has a reliability function given by

$$R(t_1, t_2) = \exp\left(-\lambda_1 t_1 - \lambda_2 t_2 - \lambda_{12} \max(t_1, t_2)\right),$$

where λ_1, λ_2, and λ_{12} are positive parameters. For most of the models it is possible to obtain parameter estimates by either maximum likelihood or Bayesian methods. Unfortunately the maximum likelihood estimates for the Marshall and Olkin model need a non-standard approach, see Proschan and Sullo (1976). The reason for this is the non-zero probability that $t_1 = t_2$, see Downton (1970).

Some of the exponential models can be easily generalized to multivariate Weibull distributions. Other formulations are possible including Weibull mixtures, see Crowder (1985). There are a number of other multivariate models which could be considered, see Clayton (1978) and Johnson and Kotz (1972). Smith (1991) explored multivariate extreme value distributions which might be applied to reliability data, in particular the estimation process. One concern was whether the estimates obtained for individual components, without taking into account the dependency, could be used as a good approximation or starting point for the estimation taking into account dependency. Not surprisingly they found that for the exponential distribution, with specific forms of dependency this was a useful approach, but generally this was not the case.

Where data are censored by the first component's failure, as in the case of a series system, then the competing risks models can be used, see Cox (1964). The first component's failure time is known and all the other components' failure times have been randomly censored. Several authors have considered the analysis of such data, see Kalbfleisch et al. (1978). There is a problem of identification since results by Cox and others indicate that for any dependent model which fits the data there exists an independent model which fits equally well.

Component

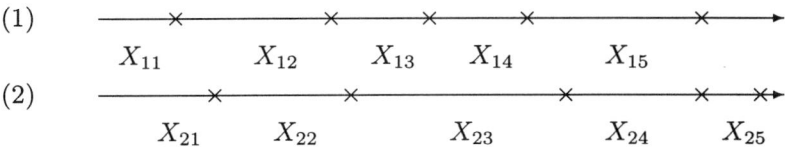

Fig. 8.3. Two processes with multivariate links × denotes an event or failure.

The models discussed so far are mainly for components although some of the exponential models are also applicable to processes. There is no fundamental problem associated with extending multivariate models to processes although difficulties can arise from defining the structures of dependency. In the example in Fig. 8.3 it is possible to contemplate a multivariate distribution for $X_{11}, X_{12}, X_{13}, X_{14}, X_{15}, X_{21}, X_{22}, X_{23}$, and X_{24}. The properties of the distribution have to be defined. Is it sensible to imply that X_{24} should have an effect on X_{11}? More sensible models might suggest that lifetimes in one process are dependent on previous lifetimes in that process and also previous lifetimes in other processes. These would need too many parameters. It could be argued that only the last observation from each process should be used. This would reduce the number of parameters significantly. There are other problems to be resolved about the method of estimation. Hence there is still a need for more work beyond that done for the multivariate exponential distributions.

A multivariate time series approach might be an appropriate alternative, see Soyer (1986).

8.5.6   RENEWAL PROCESSES

A review of renewal processes was given in Chapter 5. They can be used to model specific types of dependency. Again there is a long history of application of such models in reliability, see Cox and Lewis (1966). Downton's and Marshall and Olkin's models can be considered as a superimposition of renewal processes as illustrated in Fig. 5.3 and is given here in Fig. 8.4. Every component sequence is subject to two shock processes. The first shock process (B) is restricted to the component, the second shock process (A) is shared with other sequences. The shocks which affect the component are the superimposition of both A and B which is represented as the total shock process (C). This leads to models such as the bivariate exponential models of Downton (1970) and Marshall and Olkin (1967).

There is no requirement to assume an exponential distribution, though

Process

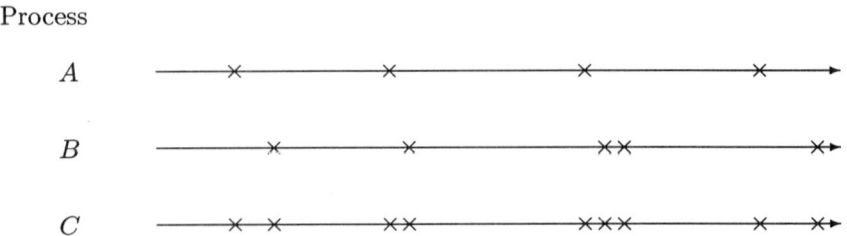

Fig. 8.4. Superimposition of shock processes × denotes an event or failure.

for simplicity many authors do. If other distributions are used then it is necessary to record not only the time since the last event, but also the time since the last joint event. Differences in the models depend on how the processes interact. If a renewal of the component process occurs when there is a renewal of the joint process then this is a semi-competing risk model. There would be difficulties in modelling if a renewal of the component process affects the joint process.

An early alternative to the superimposition renewal model (SRP) was the branching process model, see Lewis (1964, 1967). This model assumes that there are a set of initiating events, referred to as the primary events, which follow either an HPP or NHPP. These primary events give rise to subsidiary events, see Chapter 5, Subsection 5.5.2.

8.5.7 ENVIRONMENTAL MODELS

The environmental model assumes that dependency between the processes is linked through some environmental factor(s). If a factor, such as temperature or stress, is identifiable and observable then it is possible to construct the model taking this into account. A wide variety of models have been considered in which lifetimes have been related to factors. These include both parametric, see Smith (1991), and semi-parametric models, see Cox (1972) and Chapter 3. Few authors have explored the practical use of such models to describe dependency. Ansell and Phillips (1989) suggested modelling dependency through a common stress. It will be assumed that there are m series of components running in parallel. They assumed that the hazard function $\lambda_j(x, t, t_{i-1,j})$ at time t for the ith component in the jth series, which started operating at time $t_{i-1,j}$ and has been operating for a period x, could be expressed as a function of the baseline hazard function for the component $\lambda_j(x)$ and a common stress factor $G(t)$ so that

$$\lambda_j(x, t; t_{i-1,j}) = g\{\lambda_j(x), G(t)\},$$

where g is an arbitrary positive function. Choice of the appropriate model depends on belief about the context. Cox (1972) employed the multiplicative formulation for proportional hazards models but there is no reason why other models should not be chosen. This multiplicative choice of $g\{\lambda, G\} = \lambda G$ will be used here. Following Cox and Lewis (1966) the usual choice for $G(t)$ would be $G(t) = \exp(\beta t)$ or more generally $G(t) = \exp(\beta_1 t_1 + \beta_2 t_2 + \ldots + \beta_n t_n)$. Again, other models have been considered, see Etezardi-Amoli and Ciampi (1987) and Ansell and Phillips (1989).

The baseline hazard function, $\lambda_j(x)$, can be assumed either known, the parametric case, or unknown, the semi-parametric case. Ansell and Phillips (1989) used the parametric model taking the underlying distribution to be exponential when $\lambda_j(x) = \lambda_j$. Estimates can then be obtained then by maximum likelihood. A general likelihood equation is given for the exponential model in Ansell and Phillips (1989). The data consist of k_j observations of lifetimes from the jth series $x_{1,j}, x_{2,j}, \ldots, x_{k_j,j}$, with the series truncated at T. Taking $G(t)$ to be $\exp(\beta_1 t + \ldots + \beta_n t^n)$ then the equations for $\hat{\beta}_u$ and $\hat{\lambda}_j$ are given by

$$\sum_{j=1}^{m} \sum_{i=1}^{k_j} t_{ij}^u = \frac{k_j \int_0^T t^u \exp(\sum_{s=1}^{n} \hat{\beta}_s t^s) dt}{\int_0^T \exp(\sum_{s=1}^{n} \hat{\beta}_s t^s) dt}, \quad \text{for } u = 1, 2, \ldots, n,$$

and

$$\hat{\lambda}_j = \frac{k_j}{\int_0^T \exp(\sum_{s=1}^{n} \hat{\beta}_s t^s) dt}, \quad \text{for } j = 1, 2, \ldots, m.$$

Example 8.5.4
The data given in Table 8.3 are failure times (in hours) for two components working in the same environment truncated after 18 hours. The results of applying the parametric model are given in Table 8.4. The radical changes in the parameter estimates as the polynomial increases should be noted. This parallels the results of Lewis (1972) when fitting to a single population.

□

Estimates for the semi-parametric model can be found using partial likelihood, see Cox (1975). It should be noted that the covariate stress varies with time. The details of the model are given in Chapter 3 and a more extensive analysis using the model is presented in Chapter 10, Section 10.2.

Table 8.3. Failure times on two components working in the same environment

	Times of failures (in hours)								
Series 1	1.2	3.6	5.1	7.9	10.8	11.9	12.9	15.9	16.1	17.2
Series 2	0.5	6.8	7.8	10.4	10.7	11.3	11.9	12.6	15.7	

Table 8.4. Estimates for the parametric environmental model for the data in Table 8.3

	λ_1	λ_2	β_1	β_2	β_3	log(like)
Linear	0.3874	0.3487	0.0379			−30.752
Quadratic	0.2536	0.2283	0.1620	-6.56×10^{-3}		−30.535
Cubic	0.5572	0.5015	−0.3272	0.0588	-2.38×10^{-3}	−29.962

8.6 Discussion

Dependency will be a major concern in the future as systems become more complex and their function less simple to comprehend. The approaches briefly outlined in this chapter indicate how problems are or could be approached. The paucity of data often suggests the use of relatively naive methods. These methods may give misleading results. More important they may overlook some of the forms of dependency to which a system may be exposed. The statistical models provide more rigorously defined descriptions and a richer medium in which to explore dependency.

However, if there is only limited information available it might often be more appropriate to expend resources on further collection of data rather than attempt to develop a sophisticated model whose validity would surely be in doubt.

9
Practical aspects of reliability data analysis

9.1 Introduction

Previous chapters have explored models and techniques which provide the basis for reliability data analysis. In this chapter the practical aspects are discussed. Before starting on such a discussion, though, it is worth remarking that many reliability analyses do not require the use of statistical techniques. Even for those which do, the use is often very limited and routine. Sometimes their use may even be misguided, see O'Connor (1991). The statistical analysis will often not play the central role in the full analysis and will mainly be used to support the conclusions about the main objective. However, the use of statistical analysis can provide considerable information to enable analysts to make decisions. Often the value of this analysis is overlooked and the full power of the statistical tools available underutilized. One reason is that in the past reliability engineers have seen sophisticated techniques applied to their data with relatively little advantage to them. There has been a concentration on the technique rather than the underlying problem. When statistical methods are used the approach should be problem-led rather than technique-led. This means that the objective of the analysis should always be to the fore in the study.

The objective is the key to the analysis. The way to achieve the objective is through a strategy. The first section of this chapter will therefore concentrate on objectives and strategies. The strategy offered in this chapter is not prescriptive; hence unlike other authors a flowchart is not provided for the possible stages in the analysis. Obviously different objectives will have specific strategies developed for them. The strategy should be flexible so as to allow for divergence when seeking to gain insights into the processes under study. In the analysis of data there is a benefit to be gained by elementary and exploratory techniques being applied at an early stage. These should include both graphical analysis and descriptive and summary statistics. After this initial stage the approach to the analysis will be dependent on the objective. Whilst this means the use of specific techniques it is possible to discuss general issues which apply to a wide range of techniques.

In Chapter 1 the form of the data and the time scale of measurement were considered. It is important to choose the most appropriate time scale. This may be calendar time, operating time, or the number of times a component (system) has been used. It may be difficult to find the actual value of the operating time and it may be necessary to use calendar time as a proxy. The form of the data obviously depends on the context. For some situations it is only possible to obtain aggregate data on systems' or components' behaviour rather than individual lifetimes. The data may consist of single lifetimes, a sequence of lifetimes, or even multiple sequences of lifetimes. Each type will have an effect on the analysis and this will be highlighted where appropriate.

9.2   Objectives and strategies for reliability studies

Any study should have clearly stated objectives as was explained in Chapter 1, Section 1.3. It may be that some of the objectives are not wholly defined and are a little vague, such as in an exploratory study of a process, but even in that case it should be possible to declare at the beginning what the end point would be. If there is no objective then there are likely to be few positive results from the study.

Given an objective, or goal, it should form the basis of the analysis. The aim of the analysis is to satisfy the objective and the process by which this objective is achieved is the strategy. The strategy does not have to be an elaborate affair, but should be capable of implementation as a sequence of steps. The strategy should be sufficiently flexible to allow the results obtained in the process to affect the analysis. It should also allow scope for divergence so that useful insights about the process can be made.

Strategy has had limited attention in the statistical literature, until recently. An early reference to the need for strategy was Cox and Snell (1968). The usual strategy is an iterative approach, which involves the revision of hypotheses based on the analyses until sufficient comprehension is gained about the context. The hypotheses describe the previously held views about the model of the process or the distributions involved. The analysis will usually be a technique from statistical inference applied to data. The resulting fit from applying the technique will lead to model criticism which will update the hypotheses.

The techniques are applied to the data based on the hypotheses. In applying a technique to the data there are usually four stages:

(1) Selecting the technique;
(2) Application of the technique;

(3) Diagnostic tests associated with technique; and

(4) Interpretation of the results.

Selecting the technique will depend on the current hypotheses. Diagnostic tests are used to test whether the model fits the data. This can include the testing of the assumptions made about the model as well as seeing how well the model fits the data. In fact the former are often more important. Then the interpretation will decide on the current hypotheses and formulate new plausible hypotheses.

This describes the application of the technique, though it must be stressed that it is not the aim of the analysis. The technique should have been selected so that progress can be made towards the objective and the interpretation should cast light upon the objective. Often unfortunately this is not the case. In the past there has been a tendency, apparent in many reported studies, to use the latest technique independent of whether it will satisfy the objective. Many reliability objectives can fortunately be obtained by relatively simple means. The application of more sophisticated techniques, whilst providing more information, will not necessarily change the view of the objective.

Typical objectives in reliability are the prediction of future performance of a component (system) so as to guarantee a minimum lifetime, decisions on whether a component (system) has an increasing or decreasing failure rate with time, or whether one component is better than another. Each of these may be tackled at least initially by fairly simple approaches. Therefore given an objective it is always a wise step to decide clearly what would achieve the objective. It may be that in carrying out this step it is possible to see what information is required and what analysis is required. If this is not immediately clear then there possibly needs to be a clarification of the objective so that it can be achieved.

Before passing to the step of applying techniques to the data it is sensible to comprehend the nature of the data. This means not only checking and validating the data but also getting a general feel for the data. This cannot detract from the goal of any analysis since without knowledge of the data it is hard to place any reliance on the results of the analysis obtained from it. Graphical techniques are often preferred at this stage to give insights into the data. Besides graphical techniques, summary statistics can be calculated which can provide useful information. Both of these will be discussed in Section 9.3.

9.3   Elementary exploratory analysis

The aim of this stage in any analysis is to understand as far as possible the nature of the data. The approach presented here follows that of Tukey's

Table 9.1. Engine mount bracket removal times (in cycles)

10230	14530	14970+	15650	15960+	16210	16410
25220+	35850	43190	43460+	46510	47110	48200+
51570	51770	58840	73190	78220	78570	83060
85660	87690	102520+	109150	151980	156090	198150
233400	279920	290050+	313260	348060	531650	

Note: Figures with + are right censored observations, not failures.

EDA (Exploratory Data Analysis), see Tukey (1977). Simple plots and statistics should be used at the initial stage. Descriptive statistics usually calculated are the *sample mean*, the *median* (middle value), the *sample standard deviation* or *variance*, the *minimum* and the *maximum* and their difference the *range*, the *lower* and *upper quartiles*, and their difference the IQD (InterQuartile Distance). The *index of dispersion* is also a useful summary statistic. A good account of the initial examination of data was given by Chatfield (1985). In reliability studies care has to be taken when dealing with censored data, see Chapter 2, Sections 2.4, 2.5, and 2.6. For many of the above statistics modifications have been suggested to cope with censoring. Obviously in such cases great care is required in the interpretation of statistics even with appropriate modification. More detailed analyses should follow based on this initial analysis. However, the data should not be allowed to define the whole analysis; the objective must always play a significant role.

The initial stage of data analysis is inspection and validation. The data have to be checked where possible for inconsistencies and possible errors. This should include checking that observed lifetimes are usually less than the period of study, that for renewable systems the sums of lifetimes are less than the period of study, etc. Repetitions of values should be checked to see whether they are genuine repeats. Similarly associated values should be checked.

The next stage is to plot the data. It is possible to divide the approaches into two parts: elementary analysis and reliability analysis. Fortunately plotting has a long history in reliability analysis and this is covered in Section 9.4. However, it is still important to emphasize that the elementary analysis should be started with simple plots, such as dot plots, stem-and-leaf plots, histograms, and box–whisker plots. These can be very help-

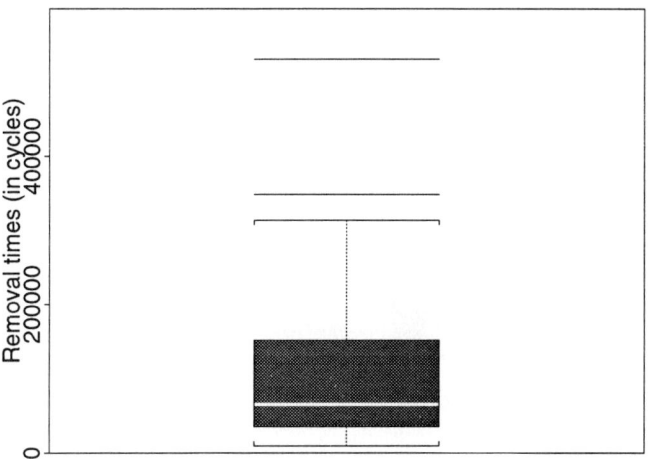

Fig. 9.1. Box–whisker plot of the engine mount brackets data given in Table 9.1

ful and even the stem-and-leaf can be very informative in an elementary analysis, as it will possibly indicate the homogeneous nature of the data. Through such plots it is possible to gain insight into the data and to see whether there are points which may be outliers or influential points.

Example 9.3.1
The use of elementary plots will first be illustrated on a data set given by Davis (1991), which has not been considered so far. These data are the removal times (in cycles) for 34 engine mount brackets and are given in Table 9.1. Fig. 9.1 presents a box–whisker plot. The horizontal line in the interior of the box is the sample median, which is 66015 cycles. The height of the box is the IQD. The whiskers (the dotted lines) extend to the extreme values of the sample or to a distance of $1.5 \times$IQD from the box, whichever is less. The two largest observations of 348060 and 531650 lie above the top whisker and hence are an indication that they are outliers. In particular the largest time of 531650 is more than twice the size of 85% of the observations, and hence may be influential if retained or excluded from the sample in any analysis.

□

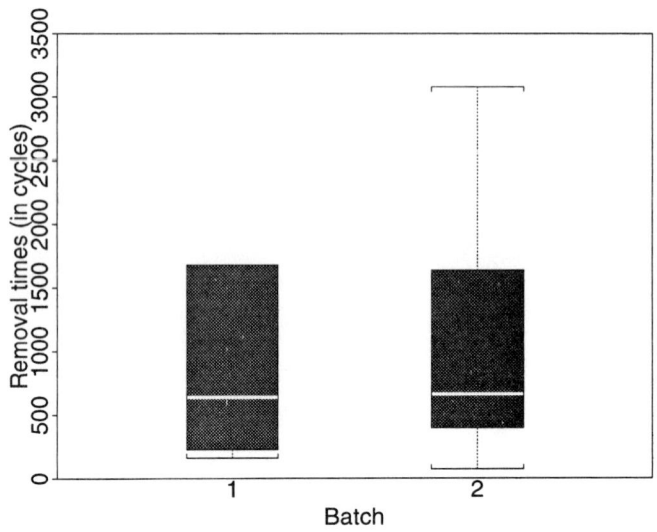

Fig. 9.2. Box–whisker plot of the sodium sulphur battery data given in Table 3.2

It is also possible to use such simple plots as part of the analysis itself. Though it is helpful to be able to use formal hypotheses tests when these tests are available, it is also helpful to be able to illustrate the conclusions of any test with a plot. For example, when comparing two or more populations it may be possible to use box–whisker plots, and the other plots, to indicate any differences between the populations.

Example 9.3.2
In Fig. 9.2 the use of the box–whisker plot is illustrated to compare two subpopulations using the data on the sodium sulphur batteries as given in Table 3.2. From this plot it is seen that both batches have approximately the same median lifetimes. However, the range of lifetimes for batch 2 is nearly twice that for batch 1, though the IQD for batch 2 is only about three-quarters of the IQD for batch 1. This demonstrates the very different behaviour in the tails (extremes) of the two batch distributions. In the plot no account is taken of the right censoring, which accounts for 4 out of the 15 values in batch 1 (the 4 largest values) as opposed to only 1 out of the 20 values in batch 2 (which is near to the median value). Again it is important to take account of censoring and thus the apparent difference between the batch lifetimes is perhaps not significant. Kimber (1990) has

suggested a correction to the box-whisker plot for censored observations from a distribution which may be skewed.

□

9.4 Exploratory reliability analysis

After these elementary plots there are plots which are more frequently associated with reliability analysis. Examples of these are hazard plots, probability plots (such as Weibull plots), plots of the ROCOF, and Duane plots. The first two are primarily used to examine the distribution of lifetimes whilst the latter are concerned with changes in the ROCOF over time, where in this latter case it is assumed that data are chronologically ordered and come from a sequence. A number of these plots have been introduced in earlier chapters so in this section they will not be repeated but instead reference will be made to the appropriate examples in the earlier chapters.

Hazard plots can be useful for assessing the shape of the hazard function or whether lifetimes have an increasing or decreasing failure rate. The cumulative hazard function, $\Lambda(t)$, can be derived from the reliability function as $\Lambda(t) = -\log(R(t))$. Hence the methods described in Chapter 2, Section 2.5, for the Kaplan–Meier (KM) or the Nelson–Altschuler (NA) estimator can be used to estimate the cumulative hazard function.

Example 9.4.1
Using the data on the removal times for the engine mount brackets in Table 9.1 a plot of the NA estimate $\hat{\Lambda}(t)$ of the cumulative hazard function against t (the removal time in kilocycles) is given in Fig. 9.3. Also upper and lower limits are given based on $1.96 \times$ the standard errors, which were obtained using the asymptotic variance estimate for the NA estimator, given in Chapter 2, Subsection 2.5.2. This can be used to assess the shape of the hazard function. There appears to be some evidence for a larger increase in the cumulative hazard function before 100 kilocycles than after this time. However, the difference does not appear to be great and this plot could be consistent with a linear function (for a Weibull distribution). It is at this point that the objective of the analysis of the removal times will play a role. The objective will decide if further consideration of this matter is important or whether the results of the present analysis are sufficient.

□

As an alternative to the hazard plot Barlow and Campo (1975) have suggested using the TTT-plot (Total Time on Test-plot). These may be helpful in deciding whether a distribution has a decreasing or increasing

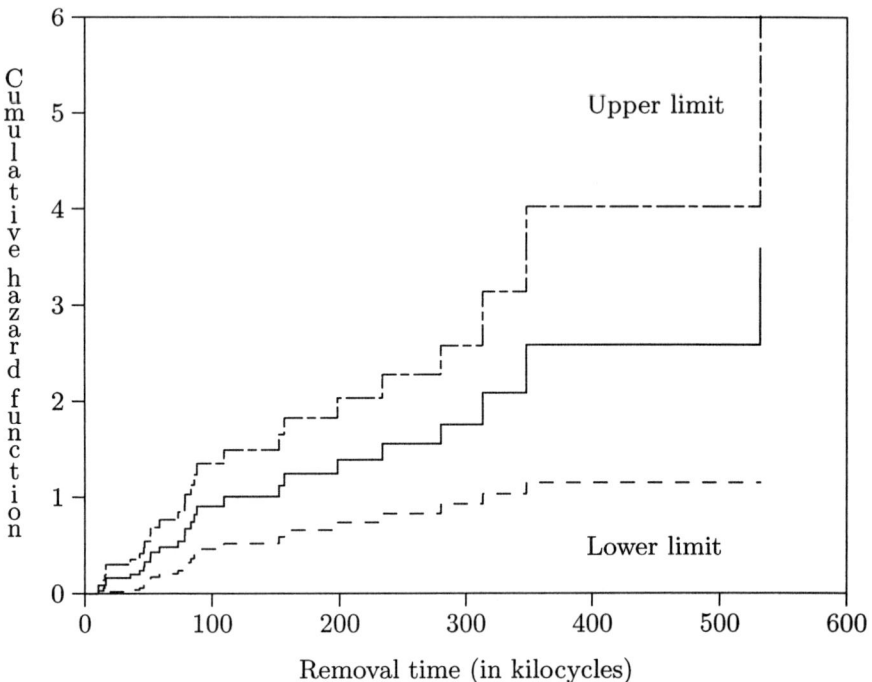

Fig. 9.3. Cumulative hazard function plot of the engine mount brackets data given in Table 9.1 with lower and upper limits

failure rate. Suppose $T_{(1)}, T_{(2)}, ..., T_{(n)}$ is an ordered sample of failure times, so that $T_{(1)} < T_{(2)} < ... < T_{(n)}$. The ith TTT statistic $T_{i,n}$ is defined by

$$T_{i,n} = T_{(1)} + T_{(2)} + ... + T_{(i-1)} + (n - i + 1)T_{(i)}$$

and the ith *scaled* TTT *statistic* $u_{(i)}$ is defined by

$$u_{(i)} = \frac{T_{i,n}}{T_{n,n}}.$$

The TTT-plot is then obtained by plotting $u_{(i)}$ against i/n. If the distribution is IFR (Increasing Failure Rate) (DFR, Decreasing Failure Rate) then the plot should be convex (concave).

There have been proposed extensions to TTT-plots, see Akersten (1987). If some of the data are censored then Barlow (1989) suggests that the plots

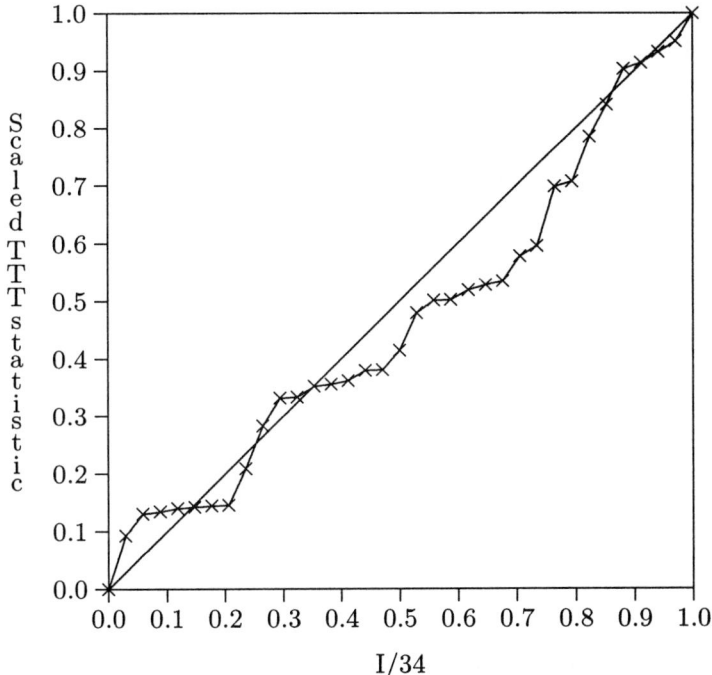

Fig. 9.4. Total time on test (TTT-)plot of the engine mount brackets data given in Table 9.1

may be misleading. Lawless (1989) has also suggested that simply using the Weibull plots can give the same information.

Example 9.4.2
Using the data on the removal times for the engine mount brackets in Table 9.1 a TTT-plot is given in Fig. 9.4. This plot suggests that the distribution of the removal times is neither IFR or DFR. So a constant failure rate distribution (the exponential distribution) may be appropriate. This will be further considered in Example 9.4.3.

□

At this stage more specific distributional forms will become appropriate to the study. So *probability plots* such as Weibull plots are useful. The probability plot chosen assumes some distributional form. The interpretation of such plots has many pitfalls. It should be remembered that the objective of the analysis must play a very significant role in interpretation.

One of the most serious pitfalls is reading too much into the plot. Certainly the departure of the extreme points from any predicted line is often not significant.

These probability plots are based on estimating the reliability function $R(t)$ or the cumulative hazard function $\Lambda(t) = -\log(R(t))$ to assess whether the sample comes from a specified distribution. These plots compare the estimated reliability (cumulative hazard) function with the theoretical reliability (cumulative hazard) function assuming that the distribution is known. When there is no censoring these plots were originally constructed using the quantiles of the distribution in the following way. The ordered lifetimes, $t_{(i)}$, were plotted against the quantiles of the distribution, $R^{-1}(1-i/n)$. These are sometimes referred to as Q–Q plots. These plots are equivalent to using the KM estimator of the reliability function for a sample of uncensored observations with no ties.

It is usual to use transformations to make the plots linear. It is assumed that it is possible to judge whether the plot is approximately linear or not, and hence to assess the distribution's fit to the data. For example, the cumulative hazard functions for the exponential and Weibull distributions are λt, from Chapter 1, Subsection 1.6.3, and λt^κ, with $\lambda = 1/\theta^\kappa$, from Chapter 2, Subsection 2.3.2, respectively. Thus a plot of $\hat{\Lambda}(t)$ against t in the case of the exponential distribution should be linear and an estimate of λ would be the slope of an appropriately fitted line. For the Weibull distribution a plot of $\log(\Lambda(t))$ against $\log t$ should be linear, since

$$\begin{aligned} \log(\Lambda(t)) &= \log \lambda + \kappa \log t \\ &= \kappa(\log t - \log \theta). \end{aligned}$$

This equation implies that the slope is κ, the shape parameter, and the intercept on the $\log t$ axis is at $t = \theta$, the scale parameter. Estimates of these parameters can be obtained from an appropriately fitted line. This plot is known as a *Weibull plot*.

Example 9.4.3
Using the data on the removal times for the engine mount brackets in Table 9.1 a Weibull plot is given in Fig. 9.5. Included with the plot is the 'best fit' straight line fitted using maximum likelihood estimation. This line can be used to obtain the maximum likelihood estimates of the parameters κ and θ of the Weibull distribution. The intercept on the $\log(\text{Cycle})$ axis is 11.86 which gives an estimate for θ of $\hat{\theta} = \exp(11.86) = 141.5$ kilocycles. The slope of the line is 1.071 which gives an estimate for κ of $\hat{\kappa} = 1.071$. This estimate differs little from 1 and suggests that an exponential distribution may be appropriate as was suggested in Example 9.4.2 by the TTT-plot in Fig. 9.4. Using the asymptotic maximum likelihood results outlined in

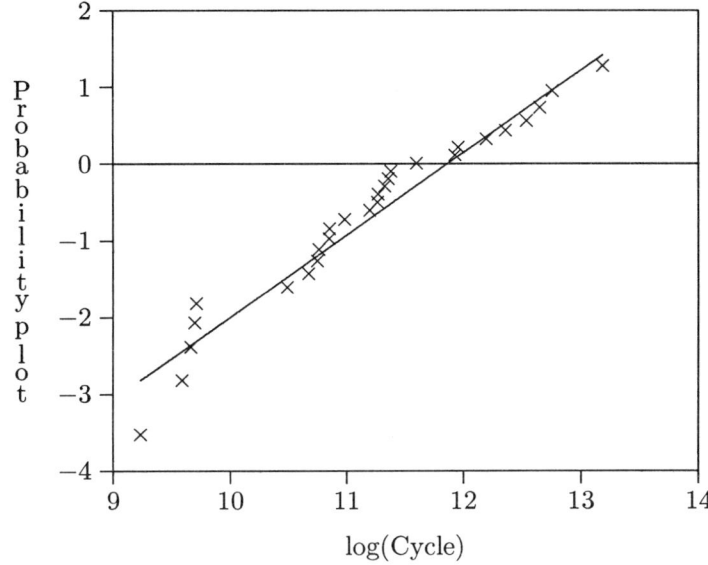

Fig. 9.5. Weibull plot of the engine mount brackets data given in Table 9.1 with 'best fit' line fitted using maximum likelihood estimation

Chapter 2, Subsection 2.6.3, which were illustrated for the Weibull distribution in Example 2.6.4, it is possible to obtain the standard error for the estimator $\hat{\kappa}$, which is 0.142. Hence a 95% confidence interval for κ would be (0.79, 1.35), which includes the value 1. A Wald test of the null hypothesis that $\kappa = 1$ would give a p-value of 0.62. Hence using an exponential distribution for the removal times is acceptable.

Apart from using the plot to estimate the parameter values of the distribution, the other purpose of the plot is to judge whether the model assumptions, which imply that the plotted points should roughly lie on a straight line, hold. This is, of course, a matter of judgement. It is important to emphasize that it is quite usual for the points on the left-hand edge of the plot (corresponding to small values of time) not to lie close to the 'best fit' straight line. So this should not cause the Weibull distribution to be rejected. As for the rest of the plotted points it could be argued that most points are above the line before $\hat{\theta} = 141.5$ kilocycles and below the line after this time. However, generally the departures from the straight line do not appear to be too great and the assumption of a Weibull (or exponential) distribution is probably acceptable.

□

Depending on the problem and the data collected it may be possible to investigate bivariate relationships. For repairable systems a bivariate scatterplot of lifetimes against lagged lifetimes may shed some light on dependency, see Chapter 6, Section 6.7. Usually in such cases a plot with a lag of one is used first and then depending on the objective the value of the lag will be increased. There will be loss of information as the lagging increases, which means that eventually the plotting becomes uninformative.

For repairable systems data the next stage might be to test for trend and dependency, see Chapter 6, Sections 6.6 and 6.7. There are a number of tests in the literature for trend, e.g. Laplace (see Cox and Lewis, 1966) and MIL-HDBK-189 (1981). Unfortunately no test provides a general defence against all types of trend. As Stephens (1989) has pointed out the Laplace test may indicate a trend when there is none if the coefficient of variation is large, e.g. the situation for the Weibull and gamma distributions with shape parameters less than unity. The objective of the study may guide one to the most appropriate test or tests. For dependency Cox and Lewis (1966) suggested that the correlation coefficients provide an adequate indication, see Chapter 6, Section 6.7.

If the data are chronologically ordered then interest can be focused on whether there are changes in the rate over time. Plots of the cumulative number of failures against the cumulative time, or cumulative operating time, can give some guidance. For example, if the process is improving times between failures should be increasing and so the curve will be convex. If the system is deteriorating then such a curve would be concave. If there are changes in the rate of the process this should also be visible on the plot. The graphical approach of Walls and Bendell (1989) detailed in Chapter 8 is based on such plots.

Another approach which is possibly easier to understand is the plot of the ROCOF (renewal density function) against time, see Chapter 6, Section 6.4. The ROCOF can be estimated by discretizing the time axis. The method is to divide the time axis into equal intervals and then divide the number of events in each interval by the interval length. This rate can then be plotted. Obviously the choice of interval size will affect the estimate to some degree, but it should be remembered that the technique is being used exploratively. If the rate does not appear constant one might contemplate using one of the models discussed in Chapter 7 and consider a plot where the form of the change is assumed known, for example in the Duane plot where there is a specified relationship between the change in the rate and time. However, such plots are more restrictive than those previously described and can be misleading.

9.5 Further reliability analysis

Beyond the exploratory stages it is harder to be specific about any further analysis since at this stage the analyst should be focusing on a specific objective. However, it is possible to discuss some general points which may aid the analyst. Assuming a given objective means that the analyst has a view about what to try to establish. It may not be wholly expressed in statistical terms but the analyst should be capable of stating what is currently known and accepted and what is plausibly true about the process. The next stage is to select the technique or techniques which will establish the objective. An analyst might, for example, wish to establish that a process is improving. The techniques described in Chapter 7 can be used as a first stage. If the analyst wishes to know whether certain factors affected the reliability of a system then the methods in Chapter 3 can be used.

Given a technique the application should be relatively straightforward. However, the next stage of deciding whether the model is appropriate is more difficult. Usually in the application of statistical methods a model is fitted by calculating some or all of the parameters of the model. There are then several areas of concern for the analyst.

Firstly are the estimates 'good'? Leaving aside the issue of whether the procedure used to estimate the parameters will produce 'good' estimates there are questions about how the data affect the estimates. There may be points which are having an undue influence on the analysis. An approach is to jack-knife the estimates, leaving one data point out, and then to re-calculate the estimates. If a point changes the estimate dramatically then it is having an undue influence and so the analyst should decide whether the point should be retained. It must be decided whether the point has arisen through errors in collection or transcription or if it might be supplying valuable information which requires further investigation. If it is an erroneous point it might be deleted. For a more detailed discussion on influential points see Barnett and Lewis (1978).

Assuming the estimates are 'good', then the next stage is to decide whether the assumption on which the model is based is valid. Often the assumption can be interpreted in terms of residuals. Do the residuals conform to the pattern that is prescribed for them? For most techniques there is advice on the appropriate tests.

Given that the assumptions seem valid the final stage is the assessment of how good a fit it is to the data. There are many ways of assessing the quality of the fit. The most usual is to compare the data with predictions for the data based on the model. This can be done on an informal basis, through plots in many cases. Alternatively it is possible to use standard goodness-of-fit tests, see D'Agostino and Stephens (1986), to indicate the goodness-of-fit of the model for the data.

9.6   Discussion

The aim of the chapter was to discuss some of the practical aspects of the analysis of reliability data which had not previously been considered. In this chapter particularly graphical approaches using plots have been highlighted. In the next chapter examples of reliability analysis will be illustrated, using the methods previously covered in the text. A general review of reliability data analysis in a brief chapter is an ambitious project.

This chapter describes approaches to data analysis rather than pursuing specific techniques. Hence the significance attached to the objective, or objectives, of a study which is often missing in statistical reviews. The chapter does not widen its scope to cover the definition of the objective as this impinges on the management of reliability studies. Other authors may feel more confident of addressing that topic. Given a clear definition of the objectives some attempt can be made to achieve them.

Given the attempt to describe an approach rather than techniques the need for practical examples is crucial. The examples in Chapters 1 to 9 and the case studies in Chapter 10 are taken mostly from the authors' experience of data analysis or from published journal papers and where possible an attempt is made to describe the context. The context, like the objective, plays a major role in the analysis. The physical context must always be used as an aid to the analysis.

The next point to be emphasized is the need for careful examination of the data before choosing a technique. The emphasis in this chapter is on simple approaches initially as basic statistics can frequently expose more insights into the reliability data than overpowerful techniques. However, when censoring is an important feature this must be taken account of in the methods used. A 'cook book' approach is not to be advocated but the following points should be highlighted:

(a) There should be a careful check (validation) of the data.

(b) Simple plots of the data should be produced.

(c) Simple statistics such as means, variances, and indices of dispersion should be computed.

(d) Trend and dependency should be investigated.

(e) Plots to investigate failure time distributions should be obtained.

(f) Finally, more specific modelling techniques such as proportional hazards modelling, multivariate analysis, or time series analysis can be used.

Before attempting (f) it is important to have spent time and resources on (a) to (e).

10
Case studies in reliability

10.1 Introduction

Before embarking on the detail of the three case studies presented in this chapter it is important to recall the role of statistical data analysis within a reliability study. However, to do this it is necessary to recall what the objectives of a reliability study might be, as was discussed in Chapter 1, Section 1.3, and Chapter 9, Section 9.2. The objectives will depend on several factors, such as who has instigated the study and why they have instigated the study. An analysis may be undertaken to establish whether a system has achieved a specific level of performance, or it may be undertaken to assess the likelihood of achieving a successful mission. The identification of where improvements can be made is also a frequent objective. This does not cover all possible aims, but it highlights the differing goals which will often imply different possible analyses. Many of the goals in reliability studies will be achieved without the need for any statistical analysis. Rarely will the statistical analysis play the central role in the study, but where statistical analysis is required then its role may be in assessment, identification, or prediction. The reliability engineer will have an initial objective and this will define the strategy for analysing the data and may also suggest the end point of the analysis.

For the case studies considered in this chapter the data consist of lifetimes which have been collected from field studies. There is an important role for laboratory-controlled data, when they are available, but a considerable amount of reliability data is from the field and this chapter concentrates on this situation. The data may be collected on a component or on a system. As well as the lifetimes there may also be associated data on other variables. This accompanying data may qualitatively describe the context or quantitatively describe covariates.

As was illustrated in Chapter 1, Section 1.4, reliability data may be censored and values may be truncated (missing). Right censoring is usually unavoidable in the analysis of field (service) data and techniques applied to these kinds of data must take this into account. It is common that a high

```
 19      0  111111111111111111111
(15)     0  222222222233333
 11      0  445
  8      0  777
  5      0  88
  3      1
  3      1 3
  2      1
  2      1
  2      1
  2      2
  2      2
  2      2
  2      2
  2      2
  2      3 0
  1      3 3
```

Fig. 10.1.   The stem-and-leaf plot of the times between the process events

percentage of the observations are censored. Three case studies are considered in this chapter, though they do not cover all the features that reliability data may possess. As far as possible for each data set the practical setting was described in Chapter 1, Section 1.2, as well as the client's objective. It is unfortunately the case for the development of the application of statistical methods to reliability analysis that commercial sensitivity is often a reason for the lack of published data.

General approaches to the analysis of failure data have been considered in Chapter 9. In this chapter the approach is more specific for each case study. Suggestions will be made for further analyses which may be carried out on the data sets. This is done for two reasons. Firstly to illustrate that statistical analysis is an iterative process. Having applied a technique to a data set it is often the case that another technique is appropriate. Secondly the objective of the analysis may have already been achieved and a desire for understanding the nature of the process for the benefit of future studies is the only reason for further analysis.

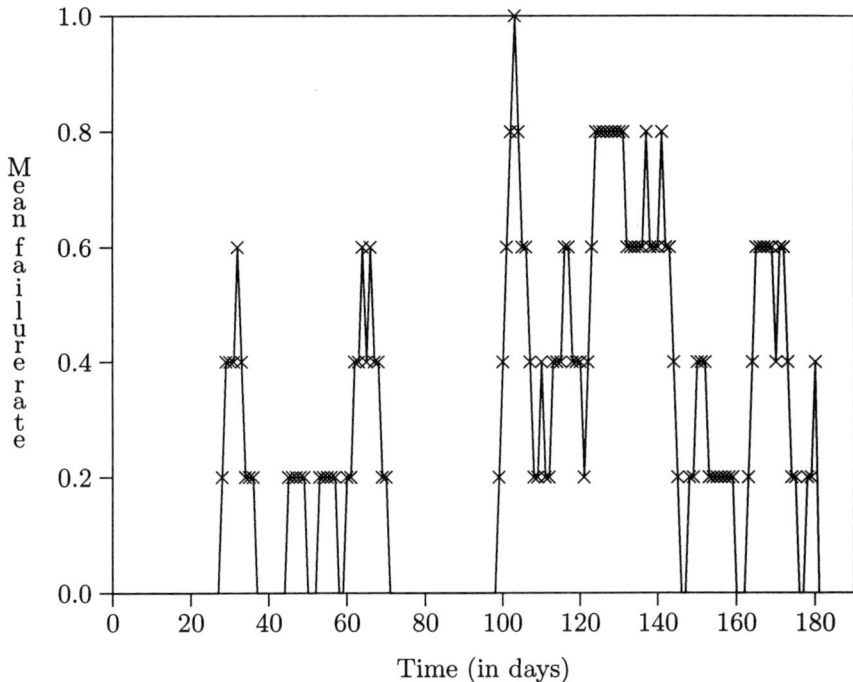

Fig. 10.2. Plot of the estimate of the ROCOF for the process event data given in Table 1.1 using the five-point moving average

10.2   Process event case study

The background to this study was described in Chapter 1 in Subsection 1.2.1 and the data (of days on which an event took place) were presented in Table 1.1. As previously indicated in Chapter 9, Section 9.3, the initial stage is to examine the data by elementary exploratory analysis. The stem-and-leaf plot for the times (days) between events (failures) is given in Fig. 10.1. The plot is positively skewed and does indicate the possibility of two 'outliers'. At such an early stage of the analysis it would seem a little unwise to eliminate these points, until further evidence is obtained. From the stem-and-leaf it is not possible to assess whether the data are under-dispersed or overdispersed compared with the exponential. The index of dispersion indicates that the data are overdispersed with the two 'outliers' and underdispersed without them. The nature of the process under study and the five-point moving average of the data given in Fig. 10.2 do indicate

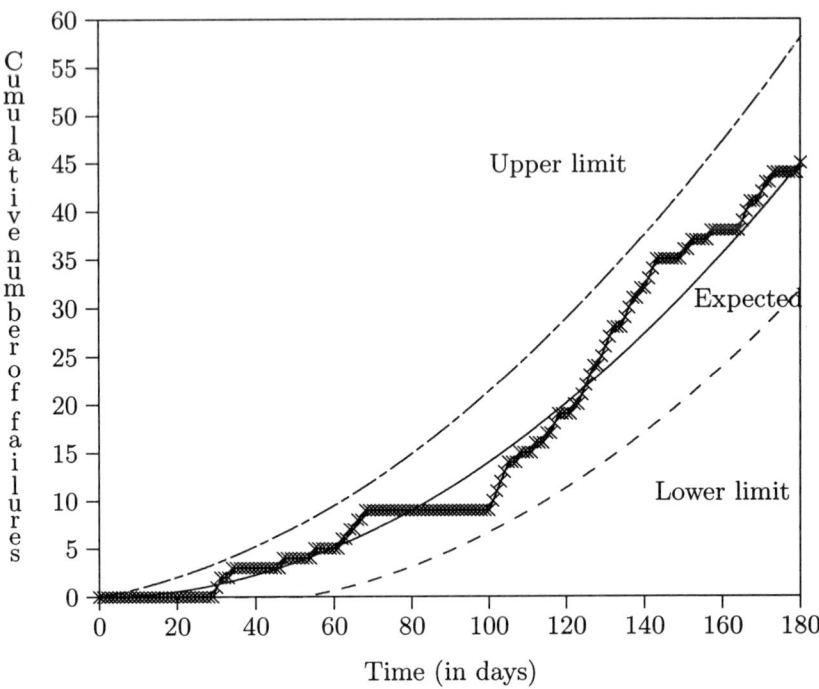

Fig. 10.3. A plot of cumulative number of events (failures) against days with expected values and confidence intervals based on Crow's NHPP model

quite clearly that there is a non-constant failure rate for the process. Calculation of the Laplace trend test statistic gives a value of -3.4 which does indicate a trend. A non-stationary model is appropriate. The result of fitting Crow's NHPP model, described in Chapter 7, Subsection 7.2.4, to the cumulative number of failures is shown in Fig. 10.3. The estimates of the parameters of the model are $\hat{\lambda} = 0.00153$ and $\hat{\beta} = 1.982$. The model does not seem to fit particularly well, though the cumulative number of failures falls within the approximate confidence intervals, described in Chapter 8, Section 8.4, and so indicates no immediate concern about dependency. The relative broadness of the confidence interval, though, suggests that there may be a lack of evidence to establish whether there is dependency or not.

The objective of the analysis is to assess the frequency of failures, and to ascertain whether these are related to the covariates or other factors. Given that there are only 45 failures in the 6 months under study an over-elaborate model would not be appropriate. Given the presence of information on temperature and stress it is possible to examine whether these covariates have an effect on failure. Both these measures vary with time and so they may account for the already established non-stationarity of the data. Of the models to incorporate covariates, described in Chapter 3, the proportional hazards model has recently been the most popularly applied model. The semi-parametric nature of the model has the appeal of not being over-restrictive. The assumption of proportionality may cause some concern, though this should not cause overdue concern with an exploratory analysis, where the objective is to discover which variables have a significant effect on failure. Solomon (1984) fortunately indicated that at least the significance of the covariates should not be affected by a choice between the proportional hazards model and the accelerated failure time model, when there is limited censoring. The results do not cover all possible departures from the model but give some reassurance in its use.

Care, though, is required when applying the proportional hazards model. The covariate data consist of daily measurements of stress (S) and temperature (T) throughout the six month observation period. The events are recorded as having occurred or not occurred on a day. Should the covariates be treated as affecting the events only on the day of occurrence or cumulatively? The first model suggests an instantaneous effect of the covariates, while the second suggests a build-up of the effect over time. The nature of the process is such that either model may be appropriate. Hence it is wise to explore both models. It is also possible to question the form of the relationship, whether it is really linear in form or of a more complex structure. Obviously in the initial analysis the use of a linear model would seem wise.

The results in Table 10.1 seem to suggest that there is an effect due to the covariates, but it is not clear which of the two is to be preferred. However, the results from the cumulative variables suggest that the stress variable is more likely to be of interest. Having established that the covariates have an effect, which achieves one objective, it is necessary to consider their incorporation in the model so as to be able to make predictions. This second objective proves to be more difficult to achieve. The proportional hazards model is a specific model and if it is to be used for prediction it is necessary to establish its appropriateness. The model assumes the proportionality of the hazard function for the covariates. This can be examined by considering the residuals derived from the model, see Kay (1977), because if the model was correct the residuals should be independent and identically distributed with the exponential distribution, see Chapter 3, Section 3.5. This is not the case for the model with both covariates as can be seen from Fig. 10.4. Hence an alternative model needs to be sought.

Even if the proportional hazards model had proved to be acceptable

Table 10.1. Results of proportional hazards analysis for the event process with covariates

Model	Coefficients (Standard errors)				Deviance
	β_S	β_T	β_{S_C}	β_{T_C}	
Null					275.62
S	3.20 (1.52)				269.88
T		−0.13 (0.06)			269.51
S+T	1.49 (2.51)	−0.09 (0.10)			269.15
S_C			−1.72 (0.38)		206.82
T_C				−0.04 (0.01)	247.12
S_C+T_C			−2.67 (0.74)	−0.10 (0.03)	192.52

The proportional hazards model has a hazard $\lambda(t; \mathbf{z})$ at time t with covariates $\mathbf{z}$ given by

$$\lambda(t; \mathbf{z}) = \lambda_0(t) \exp(\boldsymbol{\beta}^T \mathbf{z}),$$

where $\lambda_0(t)$ is the baseline hazard and $\boldsymbol{\beta}$ are the parameters to be estimated. The two covariates are denoted by S (for stress) and T (for temperature), and the respective cumulative measures are denoted by S_C and T_C.

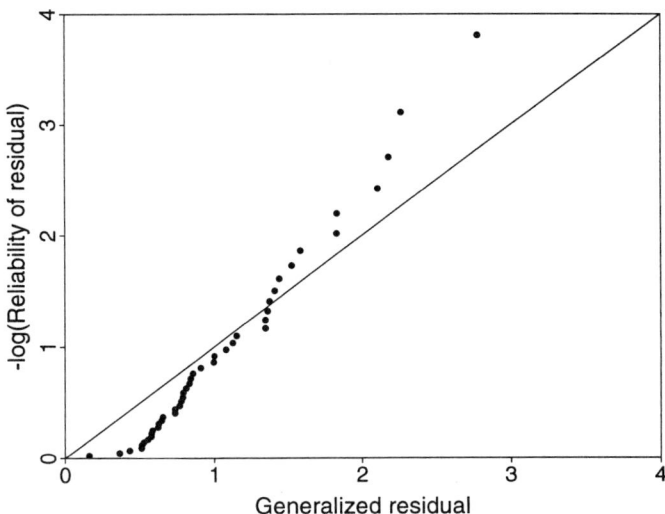

Fig. 10.4. Residual plots of the event process with covariates data for the proportional hazards model when fitting S (stress) and T (temperature)

then applying it would not be easy. One can estimate the underlying distribution function, and it may be possible then to fit a specific distribution. The next step would be to incorporate the covariates into the distribution. This may prove far from easy given their variation with time. It may be necessary to use simulation to provide the forecasts of performance.

One aspect of the data which has not been accounted for in the above exploratory analysis is the discrete nature of the data. It would be possible to explore a discrete equivalent of the model using a logistic model, as described in Chapter 3, Section 3.7.

An alternative analysis which might prove equally fruitful is to fit a common stress model to the data, as described in Chapter 8, Subsection 8.5.7. If it is established that there is a common stress then it is possible to see if the common stress can be explained in terms of the covariates. The common stress model assumed that the stress could be explained by a polynomial. The results for the log likelihood were given by Ansell and Phillips (1989) and are presented in Table 10.2. As can be seen from Table 10.2, whilst there is advantage in fitting a linear model there is no significant improvement when fitting higher order terms. The results are consistent with the result obtained for the Laplace trend test statistic and

Table 10.2. Results of using the common stress failure model for the event process with the stress covariate

Model	Log likelihood
Constant	-159.58
Linear	-149.16
Quadratic	-148.92
Cubic	-148.10
Quartic	-147.54
Quintic	-147.51

the parameters from Crow's NHPP model. It might be noted, though, that the inverse of the polynomial for the quintic was close to the pattern of temperature measurements.

The overall conclusion is that throughout the period there is an increasingly likelihood of failure. The failure can be modelled by the covariates, though there is concern over the use of the proportion hazards models because of the residual analysis. There would also be some difficulty associated with prediction.

10.3 Electronic system case study

This second study was introduced in Chapter 1, Subsection 1.2.2. The method of data collection was described in Table 1.2 and an example of the data was illustrated in Fig. 1.1. There are two possible levels of analysis: the systems level and the module level. Since the failures are occasioned by the modules it seems appropriate to start the analysis at that level.

Starting with simple plots a Pareto plot of the total number of failures of a module against module number ranked by frequency of failure seems initially most helpful, see Fig. 10.5. This immediately highlights for the management of the manufacturing company the modules which are performing badly and hence those where initially resources should be focused. Four modules contribute 50% of the failures, whereas three modules did not fail at all during the study. This may be regarded as having already achieved one aim of the study, as it suggests that if modifications are needed then the design of these four modules is crucial.

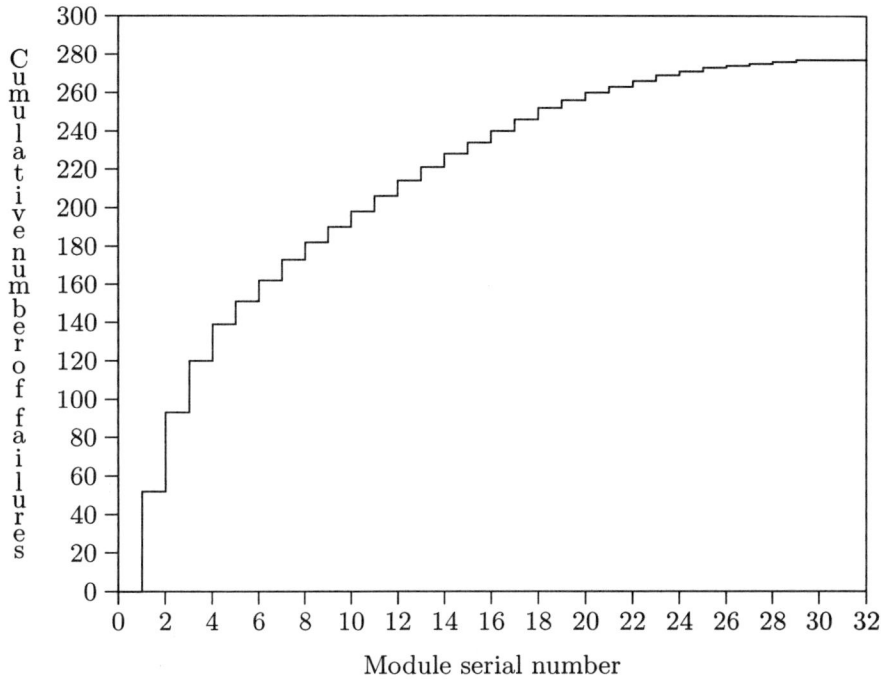

Fig. 10.5.   Cumulative number of failures for modules of the repairable electronic systems

The next stage is to see what else may be gained from the data. Obviously, initially it is best to concentrate on the modules with most failures. This would mean initially producing stem-and-leaf plots, sample means and variances, and indices of dispersion. As there is nothing of note to report in this initial analysis it is necessary to delve more deeply and to consider the distributional form and whether it is increasing failure rate (IFR) or decreasing failure rate (DFR). Could the data be treated as being from a Weibull population? Fig. 10.6 and Fig. 10.7 are the TTT-plot and the Weibull plot, respectively, of the first module. Both plots indicate strongly that the data come from a DFR distribution. The Weibull plot is similar to many empirical Weibull plots and the Cramer–von Mises statistic suggested by Koziol and Green (1976) indicates that the Weibull distribution is appropriate. Since the modules are DFR then one should replace on failure and not before.

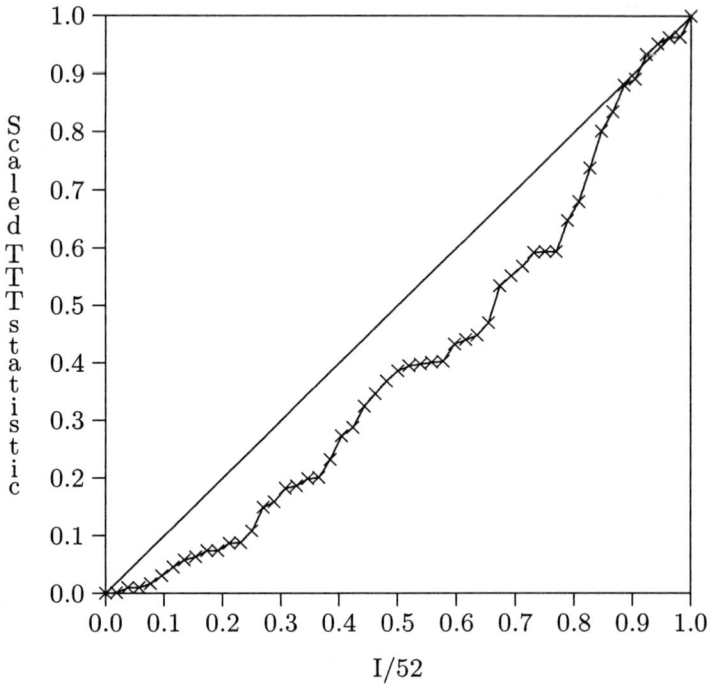

Fig. 10.6. Total time on test (TTT-)plot of failure data of module 1 of the repairable electronic systems

At this stage it is appropriate to comment again on over-interpretation. A Weibull plot for the second module was given by Ansell and Phillips (1989). It is plausible to suggest from that plot that the data illustrate the 'traditional history' of a component: an early burn-in period followed by a period of near constant failure, and then a wear-out phase. Alternatively one could suggest a non-homogeneous population. The apparent pattern might be accounted for by differing batches. Such comments cannot unfortunately be supported by the evidence available from the data. Without supportive evidence the danger arises of fitting too complex and possibly inappropriate models. If one accepts either model one would split the data into three or more groups of observations, fitting separate models to each group. Optimistic estimates may well arise from such an analysis. It is also far removed from the aim of the study.

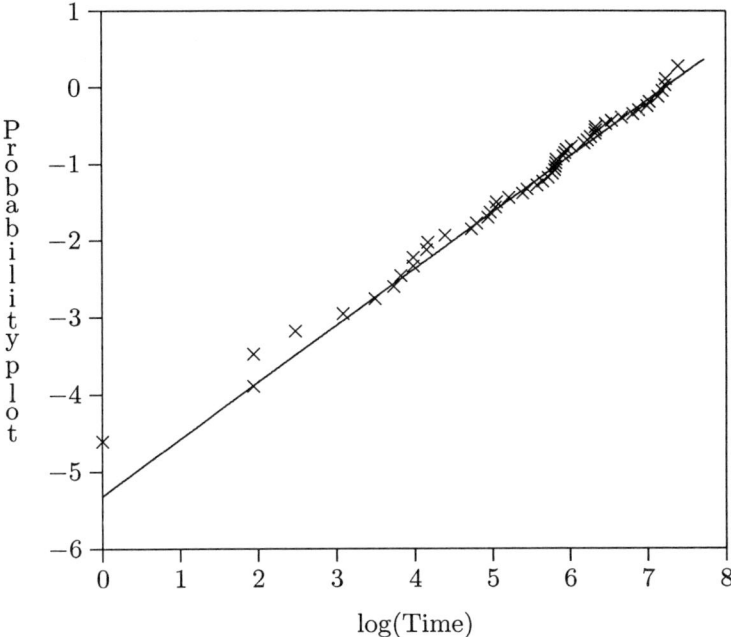

Fig. 10.7. Weibull probability plot of failure data of module 1 of the repairable electronic systems with 'best fit' line

The failure data for another 17 modules were analysed in a similar way to that outlined above for the first module. As a result of this analysis of the 18 highest failing modules the maximum likelihood estimates indicate that there is no significant difference between the failure time distributions of some of these modules. The four highest failing modules (1–4) were treated separately but the next 14 highest failing modules (5–18) were grouped into four groups. The previous analysis of failure times was repeated for these four groups and the results are given in Table 10.3.

The remaining 14 modules had four or fewer recorded failures per module. With such sparse data it did not seem reasonable to perform any elaborate model fitting. Instead these modules were assumed all to have failure times from a common negative exponential distribution. This result is also included in Table 10.3. There are two groups (A and B) of fairly reliable modules which can be fitted by negative exponential distributions with medians of about 8000 and 10000. Then there are two groups (C and

Table 10.3.   Maximum likelihood estimators of the parameters of the
Weibull and negative exponential distributions for the failure times of the
grouped modules of the repairable electronic systems

Group	Serial no. of module	Number of failures	Shape parameter	(s.e.)	Scale parameter	Median
	1	52	0.7102	(0.0852)	2753	1643
	2	41	0.8996	(0.1194)	2833	1885
	3	27	0.7564	(0.1282)	5938	3658
	4	19	0.9266	(0.1821)	6234	4198
A	7,9,14	26	1.2200	(0.2083)	7923	5868
B	8,10,13 15,16	36	0.9114	(0.1361)	18545	12404
C	5,6,12	31	0.6428	(0.1061)	34620	19574
D	11,17,18	20	0.5261	(0.1094)	170396	84901
	19–32	25	1.0		58453	40517

D) of reliable modules, which exhibit a high initial hazard rate, which can
be fitted by Weibull distributions of about 0.6 and 0.5 respectively and
with medians of about 20000 and 85000.

The analysis of these failure data illustrates some practical problems.
Firstly any system of electronic modules will typically contain many modules and some of these will have few failures. In practice, to do any parameter estimation it is necessary to group together similar modules. There
are some dangers in doing this on the basis of the data rather than for prior
physical/engineering reasons. Secondly there is the problem of highly censored data. Any plotting methods used for goodness-of-fit must take this
into account. Thirdly it is possible to obtain estimates of the parameters of
Weibull distributions easily but obtaining standard errors of the estimates
is not so straightforward.

If this model fitting is acceptable the final problem is obtaining the failure rate for each module and combining them to obtain the failure rate of
the system assuming that the modules operate independently. The failure
rate is far from easy to obtain for certain distributions. Baxter *et al.* (1981)
have produced tables for the Weibull distribution but these unfortunately
are incomplete and not easily accessible. Hence for many distributions,

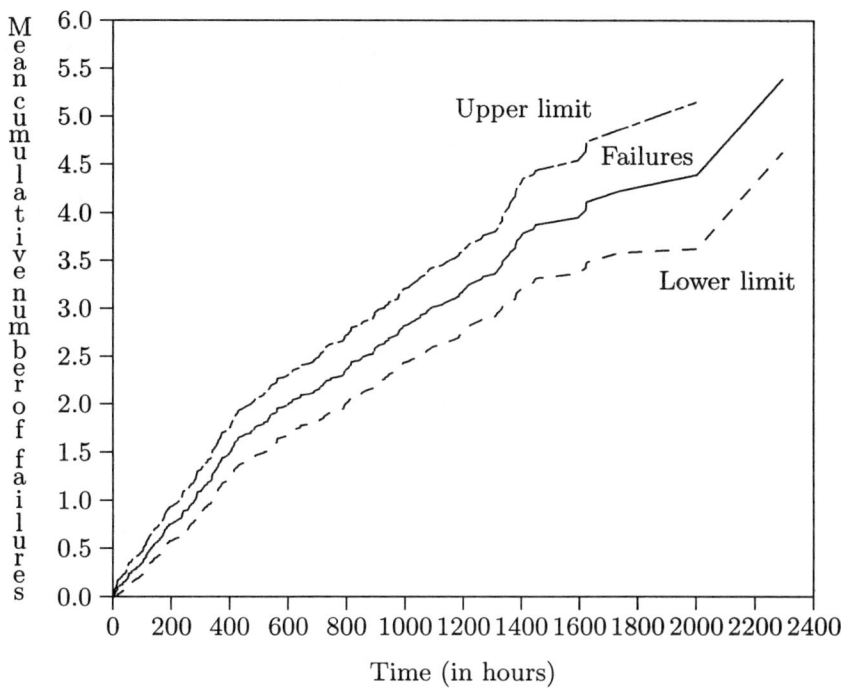

Fig. 10.8. Mean cumulative number of failures for the electronic data

using asymptotic formulae, see Cox (1962), which may not be sufficiently accurate, is the only possibility for a parametric approach. This was done by Ansell and Phillips (1989). A SRP model for the system was assumed and hence summing over the five negative exponential distributions gave a failure rate of 0.0014, i.e. 1.4 failures per 1000 hours. For the four Weibull distributions the expected cumulative number of failures using the asymptotic formula was given by

$$H(t) \sim 0.0005t + 8.4.$$

So there was a failure rate (ROCOF) of 0.0005, i.e. 0.5 failures per 1000 hours, plus 8.4 failures due to early failures because of the form of the Weibull distribution. Combining the two failure rates the total expected cumulative number of failures for the electronic system was given by

$$H(t) \sim 0.0019t + 8.4.$$

So the asymptotic failure rate (ROCOF) for the system was 1.9 failures per 1000 hours.

An alternative is to use a non-parametric approach, see Chapter 6, Section 6.5. Nelson (1988) suggested plotting the mean cumulative number of failures. The plot for the electronic data is given in Fig. 10.8 with upper and lower limits. These have been obtained by using a method of obtaining a non-parametric equivalent of the Kaplan–Meier estimator and Greenwood's variance estimator for recurrence data suggested by Nelson. The limits are calculated by adding and subtracting twice the standard errors. This plot suggests that the failure rate of 1.9 per 1000 hours is about right but that the 8.4 failures due to early failures is much too large (pessimistic).

This analysis illustrates that there is the problem of estimating the failure rate as it is difficult for the Weibull distribution. The usefulness of the asymptotic formulae applied by Ansell and Phillips (1989) depends on whether they can be used for reasonably small t or not. As they have been obtained by considering different failure rates and the rate of convergence varies over the different functions it is difficult in any application to be sure. But until methods of easily evaluating the failure rate are readily available this seems to be the best that can be done using a parametric approach. The alternative is to use a non-parametric approach. The analysis presented here has provided an estimate of the ROCOF for the electronic system, from which it is possible to decide on whether the performance of the system satisfies specification. If not, then the analysis also suggests a strategy for making modifications to the modules.

10.4   Fleet mechanical equipment case study

This final study was introduced in Chapter 1 in Subsection 1.2.3. The method of data collection was illustrated in Fig. 1.2 and the data are given in Table 10.4. These data illustrate that often after the initial analysis a very specific form of analysis should be chosen. In this case a specific stochastic model is chosen because of the nature of the data and also the model seemed to be plausible within the physical context. This model is used to provide an estimate of the failure (renewal) rate, which can be used to decide on the future repair (replacement) policy.

The model chosen for the analysis is a birth–immigration process model. This model is an example of the Markovian models considered in Chapter 5, Subsection 5.5.6; see Cox and Miller (1965) for further details. This model has two parameters: α, the rate of failures due to *external causes* (the immigration rate); and β, the rate of failures due to *wearing out* (the birth rate). For this birth–immigration process model the MLEs (Maximum

Table 10.4.   Number of renewals of auxiliary generators installed on a
marine vessel

Interval number (i)	Time interval (in years)	Renewals
1	1.5, 2.5	4
2	2.5, 3.5	5
3	3.5, 4.5	4
4	4.5, 5.5	2
5	5.5, 6.5	4
6	6.5, 7.5	11
7	7.5, 8.5	19
8	8.5, 9.5	10
9	9.5, 10.33	14

Likelihood Estimators) $\hat{\alpha}$ and $\hat{\beta}$ were found by Triner and Phillips (1986),
who obtained $\hat{\alpha} = 0.5391$ and $\hat{\beta} = 0.2117$.

The objective of the study is to enable decisions to be made about the
repair (or replacement) policy. For this reason it is necessary to have an
estimate of the failure rate (ROCOF) $h(t)$ and using the above MLEs the
estimated failure rate (ROCOF) is given by

$$\hat{h}(t) \;=\; 2.1564 \exp(0.2117t).$$

This estimate of the failure rate (ROCOF) is plotted for t in the interval
[0, 12] in Fig. 10.9. The distributions of the MLEs $\hat{\alpha}$ and $\hat{\beta}$ were investigated by simulation of the birth–immigration process with parameters
$\alpha = 0.5391$ and $\beta = 0.2117$, the observed values of the MLEs. The results
of simulations indicate that though $\hat{\beta}$ is likely to be an unbiased estimator $\hat{\alpha}$ is likely to be a biased estimator. The two MLEs have a negative
correlation coefficient.

An approximation for the standard deviation of $\log(\hat{h}(t))$ as a function
of α and β was obtained in terms of the variances $\text{Var}(\hat{\alpha})$ and $\text{Var}(\hat{\beta})$ of $\hat{\alpha}$
and $\hat{\beta}$, respectively, and the covariance $\text{Cov}(\hat{\alpha}, \hat{\beta})$, of $\hat{\alpha}$ and $\hat{\beta}$. Ignoring any
effect of bias for $\hat{\alpha}$ and $\hat{\beta}$, upper and lower limits are given in Fig. 10.9 for
$h(t)$. This is done by using $\log(\hat{h}(t))$ and adding and subtracting twice the

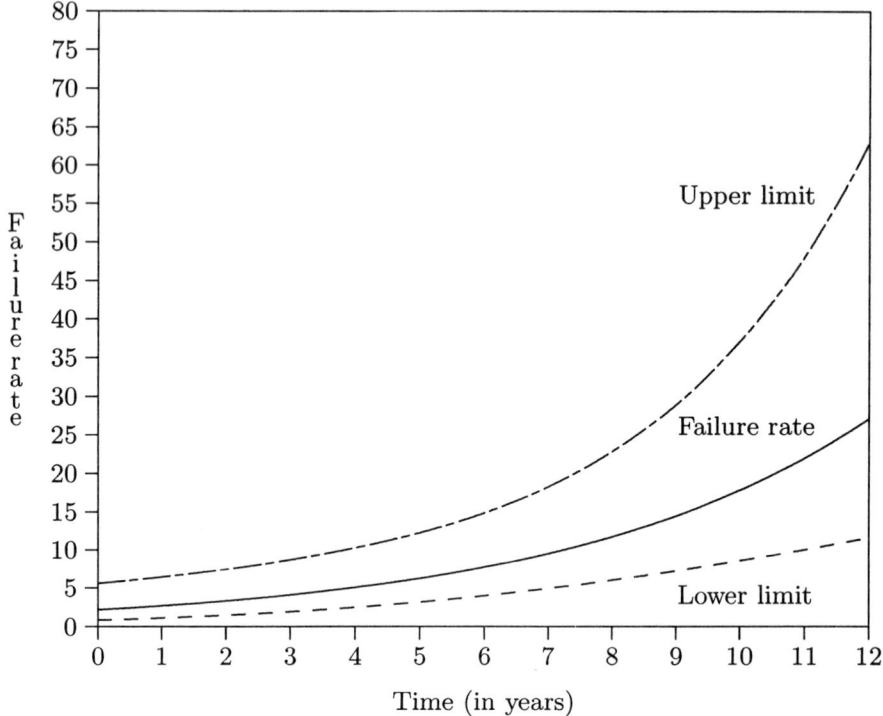

Fig. 10.9. Estimated failure rate for the aggregate data from the mechanical equipment fitted to a fleet

standard error of $\log(\hat{h}(t))$ using the simulated values obtained by Triner and Phillips (1986). This shows a slowly increasing interval between the limits up to about time 6 which then rapidly increases after time 9. Hence an optimum time for replacement would possibly lie in this interval. Economic considerations would need to be taken into account in this decision.

Estimates of the variances of MLEs can usually be obtained using standard methods from the information matrix. However, for the birth–immigration process model the usual condition of independence of observations does not apply. However, Sweeting (1989, 1992) indicated that the use of the information matrix may still be justified because of alternative 'asymptotics'.

10.5   Discussion

This chapter illustrated approaches to statistical data analysis through three case studies. These case studies considered lifetimes which have been collected from field studies. A considerable amount of reliability data is from the field and this chapter concentrated on these cases. As well as the lifetimes there were associated data on other variables (covariates).

The first study was concerned with the days on which events (failures) took place for an industrial process. The objective of the analysis was to assess the frequency of the failures, and to ascertain whether these were related to two covariates or other factors. A number of models were considered in an attempt to achieve this objective. These included the proportional hazards model, which incorporated the covariates into the model. One aspect of the data which had not been accounted for in the exploratory analysis was the discrete nature of the data, which could be explored using a logistic model. An alternative analysis which was also considered was to fit a common stress model to the data. The results obtained were consistent with the result obtained for the exploratory analysis which used the Laplace trend test statistic and the parameters from Crow's NHPP model.

The second study was concerned with the assessment of the performance of repairable electronic equipment. There was concern about the performance of the systems as well as the modules (subsystems) making up the system. The analysis illustrated that there was a problem in estimating the failure rate for a module as it is difficult for the Weibull distribution. The usefulness of the asymptotic formulae applied depended on whether they could be used for reasonably small intervals of time or not. As they have been obtained by considering different failure rates and the rate of convergence varied over the different functions, it appeared by comparison with the alternative non-parametric approach that the asymptotic formulae were not wholly appropriate. The analysis presented provided an estimate of the ROCOF for the electronic system, from which it was possible to decide on whether the performance of the system satisfied specification. The analysis also suggested a strategy for making modifications to the modules.

The third study was again concerned with a repairable system, though the system was a piece of mechanical equipment. Because of the inherent 'wear' in the system a specific stochastic model was chosen, which seemed to be plausible within the physical context. The model chosen for the analysis was a birth–immigration process model, which is an example of a Markovian model. This model has been widely studied and using the known results of this model it was possible to provide an estimate of the failure (renewal) rate, which could be used to decide on the future repair (replacement) policy. Interestingly the problem of finding the standard errors for these estimates did involve the use of recently published research work in the area of asymptotic estimation theory.

The significance attached to the objective, or objectives, of a study is central to this book, a feature which is often missing in the statistical literature. Given the attempt to describe an approach rather than techniques, the need for practical examples is crucial. The examples of these case studies are taken from the authors' experience of data analysis and an attempt was made to describe the context. The context, like the objective, plays a major role in the analysis. The physical context must always be used as an aid to analysis. Given a clear definition of the objectives an attempt can be made to achieve them and these three case studies have been used to illustrate this important message. It is necessary to remember that statistical data analysis is solely a tool to achieve a given end, though with today's powerful computing equipment it can be a very powerful tool and will become even more so in the future. The objective of any analysis is not the statistical analysis but that of the reliability study.

In Chapter 2 the approach using Bayesian methods was briefly introduced. This method has not been pursued in this book. This has not been done because of any antipathy to this approach as its value is evident. However, in a *practical* book statistical approaches which can be easily implemented with the commonly available statistical software must be presented. However, if this book is revised in the future and by then Bayesian statistical software is more widely available then this book would need to be revised accordingly. But until this becomes the case the statistical methods presented in this book are those dictated by the need for the presentation of a practical approach to statistical methods for those who are not professional statisticians.

In summing up, the approach in this book to the use of statistical analysis in reliability studies needs to be explained. The approach to estimation that has been adopted has been a mixture of the *maximum likelihood* and *classical methods*. This means that the statistical analysis has mainly concentrated on the *estimation* of a *model parameter* of interest and the calculation of a *standard error*. This is achieved by using the method of maximum likelihood estimation to obtain an *estimator* and appealing to the *asymptotic* results for large sample sizes to obtain a *standard error* for this estimator. These can then be combined to produce a *confidence interval* (which usually has 95% confidence). The interval can then be used to see if the sample is consistent with the *hypothesized* values of the parameter of interest, by checking if the interval contains this value. Alternatively a *hypothesis* test (such as a *Wald* test) can be used to obtain a 'p-value' to infer whether the sample is consistent with the *null hypothesis*. Hence the approach to statistical analysis is based on a *parameter estimator*, a *standard error* for this estimator, and a 'p-value' for any hypothesis of interest.

Appendix

To do any practical analysis of failure data from reliability studies it is necessary to use statistical software. It is beyond the scope of this book to give a guide or review of the extensive software available for general statistical analysis or for specialist reliability applications. However, this book has presented various statistical analyses using well-known and widely available software packages such as SAS and S-PLUS. For this reason it was felt that many readers might find it helpful to have examples of the programs used with these two software packages. Hence examples of an S-PLUS program and an SAS program are presented.

An S-PLUS program

The code presented below will enable a PostScript file to be produced which will give Fig. 3.7.

```
# S-plus commands for creating Fig. 3.7
#
# All comments following a # are ignored
#
# Read in the data in Table 3.2 from the file ex331.dat
# into the matrix bat.mat
bat.mat <-matrix(scan("ex331.dat"),byrow=T,ncol=3)
#
# The data were in three columns: time,
# censoring indicator, batch variable.
#
# Put these three variables into the columns time,
# status and batch.
time <- bat.mat[ ,1]
status <- bat.mat[ ,2]
batch <- bat.mat[ ,3]
#
# Fit a Cox proportional hazards model as in Example 3.5.1
# using the covariate batch and calculate the Schoenfeld
# residuals
```

```
coxreg.bat <- coxreg(time,status,batch,resid="schoen")
#
# Perform a test for linear trend in the residuals;
# Z.ph gives the p-value.
corel <- cor(rank(coxreg.bat$time),coxreg.bat$resid)
n <- length(coxreg.bat$time)
Z.ph <- 0.5*log((1+corel)/(1- corel))*sqrt(n - 3)
#
# Create the plot of the Schoenfeld residuals given
# in Fig. 3.7.
# First define suitable margins
par(mar=c(6,6,3,3))
# Plot the residuals against the lifetimes
plot(coxreg.bat$time,coxreg.bat$resid,
# main=paste("Schoenfeld Residual Plot"),
# The above line if included would give the plot a title
xlab = " ",
ylab = " ",xaxs="s",yaxs="s",
xlim=c(0,3500),
ylim=c(-0.6,0.6),cex=1.3)
# Add text to the two axes
mtext("Schoenfeld residuals",side=2,cex=1.4,line=4)
mtext("Lifetime (in cycles)",side=1,cex=1.4,line=4)
# Add the zero line
abline(h=0)
# Add the non-parametric estimate of the regression line
lines(lowess(coxreg.bat$time, coxreg.bat$resid,f=1/1),lty=2)
# Add a legend for the two lines
legend(2000,-0.4,c("Lowess smooth"," " ,"Zero line"),
lty =c(2,0,1),bty="n",cex=1.3)
```

An SAS program

The code presented below will enable a PostScript file to be produced which will create a plot of the martingale residuals against the linear predictors for the sodium sulphur battery data in Table 3.2.

```
* SAS commands for creating a plot of the martingale
  residuals against the linear predictors;
* All comments contained on a command line
* beginning with * are ignored;
options ls=78;
```

```
proc format;
value batch 0='Group 1' 1='Group 2';
value state 1='failed' 0='censored';
* Read in data from file ex331.dat ;
filename geh 'ex331.dat';
data gehan;
infile geh;
* Variables read in are time, status and batch;
input time state batch;
format batch batch. state state.;
* ;
* Sort the data by batch;
proc sort data=gehan;
by batch;
run;
* ;
* The next proc phreg produces the output dataset
  (kmout) which drives the graphics;
proc phreg data=gehan noprint;
model time*state(0) = batch;
output out=kmout xbeta =xb resmart=mart resdev=dev;
run;
* ;
* This is the graphics file which gets plotted;
filename gout 'ex331z.ps';
* Specify options for the production of the file;
goptions reset=all
 dev=psepsf
 nodisplay
 gsfmode=replace
 gsfname=gout;
* ;
* Add a title to the plot;
title1 u=1 ls=1 j=c h=1.5 F=swissx
'SODIUM SULPHUR BATTERY DATA';
* ;
data kmout;
set kmout;
run;
* ;
* The next proc gplot produces the plot;
proc gplot data=kmout;
axis1 label=(f=swissx h=1.5 r=0 a=90 'Martingale residual')
```

```
order= -2.0 to 1.0 by 0.5
value=(f=swissx h=1.5)
length=50 pct;
axis2 label=(f=swissx h=1.5 'Linear predictor')
order= -0.1 to 0.0 by 0.01
value=(f=swissx h=1.5)
length=50 pct;
plot  mart*xb /vaxis=axis1 haxis=axis2;
      symbol1 value=circle;
run;
```

These examples will hopefully be helpful for beginners using the software packages SAS and S-PLUS. However, a few remarks about these two software packages may be helpful. Both can be used interactively as well as by executing code as given in the previous sections. Data management is good in SAS, though plots are easier to produce in S-PLUS. Both SAS and S-PLUS give all necessary non-parametric analyses.

S-PLUS offers a library of functions which enable a range of survival time models to be fitted with factors and interactions. With SAS it is necessary to generate the factors and interactions before including them in the model.

Both SAS and S-PLUS give estimates of the survivor function. Also both SAS and S-PLUS provide residual plots for martingale and deviance residuals but S-PLUS also provides score and Schoenfeld residuals.

Collett (1994) gives an account of the relative merits of SAS and two other packages, BMDP and SPSS. However, for any extensive use of the software it will be necessary to study the appropriate user manuals.

References

Abramowitz, M. and Stegun, I.A. (1970). *Handbook of Mathematical Functions*. Dover, New York, NY.

Agrawal, A. and Barlow, R.E. (1984). A survey of network reliability and domination theory. *Oper. Res.*, **32**, 478–492.

Agrawal, A. and Satyanarayana, A. (1984). An $O(|E|)$ time algorithm for computing the reliability of a class of directed networks. *Oper. Res.*, **32**, 493–515.

Aitkin, M. and Clayton, D.G. (1980). The fitting of exponential, Weibull and extreme value distributions to complex censored survival data using GLIM. *Appl. Stat.*, **29**, 156–163.

Akersten, P.A. (1987). The double TTT-plot — a tool for the study of nonconstant failure intensities. In *Proc. 6th Natl. Reliab. Conf., Birmingham*, 2B/3/1–8. National Centre of Systems Reliability, UKAEA, Warrington.

Amendola, A. (1988). Common cause failure analysis in reliability and risk assessment. In *Reliability Engineering* (ed. A. Amendola and A.S. De Bustamante), 221–256. Kluwer Academic, London.

Andersen, P.K. and Gill, R.D. (1982). Cox's regression model for counting processes: a large sample study. *Ann. Stat.*, **10**, 1100–1120.

Anderson, J.A. and Senthilselvan, A. (1982). A two step regression model for hazard functions, *Appl. Stat.*, **31**, 44–51.

Andrews, J.D. and Moss, T.R. (1993). *Reliability and Risk Assessment*. Longman, Harlow.

Ansell, J.I. (1987). Analysis of lifetime data with covariates. In *Proc. 6th Natl. Reliab. Conf., Birmingham*, 2B/R1/1–14. National Centre of Systems Reliability, UKAEA, Warrington.

Ansell, J.I. and Bendell, A. (1982). On the optimality of k-out-of-n:G systems. *IEEE Trans. Reliab.*, **31**, 206–210.

Ansell, J.I., Bendell, A., and Humble, S. (1984). Age replacement under alternative cost criteria. *Manage. Sci.*, **30**, 358–367.

Ansell, J.I. and Phillips, M.J. (1989). Practical problems in the statistical analysis of reliability data (with Discussion). *Appl. Stat.*, **38**, 205–247.

Ansell, J.I. and Phillips, M.J. (1990). Practical reliability data analysis. *Reliab. Eng.*, **28**, 337–356.

Ansell, J.I. and Walls, L.A. (1990). Dependency modelling. In *Proc. 11th Adv. Reliab. Tech. Symp., Liverpool* (ed. P. Comer), 111–124. Elsevier Applied Science, London.

Ansell, J.I. and Walls, L.A. (1992). Dependency analysis in reliability studies. *IMA J. Math. Appl. Bus. Ind.*, **3**, 333–348.

Ansell, R.O. and Ansell, J.I. (1987). Modelling the reliability of sodium sulphur cells. *Reliab. Eng.*, **17**, 127–137.

Apostolakis, G. and Moineni, P. (1987). The foundations of models of dependence in probability safety assessment. *Reliab. Eng.*, **18**, 177–195.

Ascher, H. and Feingold, H. (1984). *Repairable Systems Reliability*. Marcel Dekker, New York, NY.

Atwood, C.L. (1986). The binomial failure rate common cause model. *Technometrics*, **28**, 139–148.

Aven, T. and Bergman, B. (1986). Optimum replacement times — a general set-up. *J. Appl. Probab.*, **23**, 432–442.

Ballard, G. (1989). Dependent failure analysis in PSA. In *Proc. IAEA Int. Conf. Nucl. Power Performance and Saf., Vienna*, 119–133. IAEA, Vienna.

Barlow, R.E. (1989). Contribution to the Discussion of Ansell and Phillips (1989). *Appl. Stat.*, **38**, 239.

Barlow, R.E. and Campo, R.E. (1975). Total time on test processes and application to failure data analysis. In *Reliability and Fault Tree Analysis* (ed. R.E. Barlow, J.B. Fussel, and N.D. Singpurwalla), 451–481. SIAM, Philadelphia, PA.

Barlow, R.E. and Hunter, L.C. (1960a). Criteria for determining optimum redundancy. *IRE Trans. Reliab. Qual. Control*, **9**, 73–77.

Barlow, R.E. and Hunter, L.C. (1960b). Optimum preventive maintenance policies. *Oper. Res.*, **8**, 90–100.

Barlow, R.E., Hunter, L.C., and Proschan, F. (1963). Optimum redundancy when components are subject to two kinds of failure. *J. Soc. Ind. Appl. Math.*, **11**, 64–73.

Barlow, R.E. and Proschan, F. (1965). *Mathematical Theory of Reliability*. John Wiley, New York, NY.

Barlow, R.E. and Proschan, F. (1975). *Statistical Theory of Reliability and Life Testing: Probability Models*. Holt, Reinhart, and Winston, New York, NY.

Barnett, V.D. and Lewis, T. (1978). *Outliers in Statistical Data*. John Wiley, Chichester.

Basu, A.P. and Rigda, S.E. (1986). Examples of parametric empirical Bayes methods for the estimation of failure processes for repairable systems. In *Reliability and Quality Control* (ed. A.P. Basu), 47–55. North-Holland, Amsterdam.

Baxter, L.A., Scheuer, E.M., McConalogue, D.J., and Blischke, W.R. (1981). Renewal Tables: Tables of Functions arising in Renewal Theory. *Technical Report*, University of S. California, Los Angeles, CA.

Beaumont, G.P. (1983). *Introductory Applied Probability*. Ellis Horwood, Chichester.

Bendell, A. and Walls, L.A. (1985). Exploring reliability data. *Qual. Reliab. Eng. Int.*, **1**, 37–51.

Bendell, A. and Wightman, D.M. (1985). The practical application of proportional hazards modelling. In *Proc. 5th Natl. Reliab. Conf., Birmingham*, 2B/3/1–16. National Centre of Systems Reliability, UKAEA, Warrington.

Ben-Dov, Y. (1980). Optimal reliability design of k-out-of-n systems subject to two kinds of failure. *J. Oper. Res. Soc.*, **31**, 743–748.

Billington, R. and Allan, R.N. (1983). *Reliability Evaluation of Engineering Systems: Concepts and Techniques*. Pitman Books, London.

Birnbaum, Z.W., Esary, J.D., and Saunders, S.C. (1961). Multi–component systems and structures and their reliability. *Technometrics*, **3**, 55–77.

Blumenthal, S.B., Greenwood, J.A., and Herbach, L.A. (1973). The transient behaviour of series systems or superimposed renewal processes. *Technometrics*, **15**, 255–269.

Bourne, A.J., Edwards, G.T., Hunns, D.M., Poulter, D.R., and Watson, I.A. (1981). Defences against common-mode failures: a guide for designers and operators. *Report SRD-R-196*, National Centre of Systems Reliability, UKAEA, Warrington.

Braun, H. and Paine, J.M. (1977). A comparative study of models for reliability growth. *Technical Report, No 126, Series 2*, Department of Statistics, Princeton University, Princeton, NJ.

Breslow, N.E. (1974). Covariate analysis of censored survival data. *Biometrics*, **30**, 89–99.

Bueno, J.C.C. and Sherwin, D.J. (1990). Common cause failure analysis of subsea systems. In *Proc. 11th Adv. Reliab. Tech. Symp., Liverpool* (ed. P. Comer), 125–138. Elsevier Applied Science, London.

Chatfield, C. (1985). The initial examination of data (with Discussion). *J. R. Stat. Soc. A*, **148**, 214–253.

Christer, A.H. (1978). Refined asymptotic costs for renewal reward processes. *J. Oper. Res. Soc.*, **29**, 577–584.

Cinlar, E. (1975). *Introduction to Stochastic Processes*. Prentice Hall, Englewood Cliffs, NJ.

Clayton, D.G. (1978). A model for association in bivariate lifetables and its application in an epidemiology study of fantail tendency in chronic diseases incidence. *Biometrika*, **65**, 141–151.

Cleveland, W.S. (1979). Robust locally weighted regression and smoothing scatterplots. *J. Am. Stat. Assoc.*, **74**, 829–836.

Cohen, A.C. (1965). Maximum likelihood estimation in the Weibull distribution based on complete and on censored samples. *Technometrics*, **7**, 579–588.

Collett, D. (1994). *Modelling Survival Data in Medical Research*. Chapman and Hall, London.

Cook, R.D. and Weisberg, S. (1982). *Residuals and Influence in Regression*. Chapman and Hall, London.

Cosgrove, A. (1992). A study of component dependency using a shock model. *Unpublished dissertation*, Dept. of Computer Science, University of Edinburgh.

Cox, D.R. (1962). *Renewal Theory*. Methuen, London.

Cox, D.R. (1964). Some applications of exponential ordered scores. *J. R. Stat. Soc. B*, **26**, 103–110.

Cox, D.R. (1972). Regression models and life tables (with Discussion). *J. R. Stat. Soc. B*, **34**, 187–220.

Cox, D.R. (1975). Partial likelihood. *Biometrika*, **62**, 269–276.

Cox, D.R. and Lewis, P.A.W. (1966). *The Statistical Analysis of Series of Events*. Methuen, London.

Cox, D.R. and Miller, H.D. (1965). *The Theory of Stochastic Processes*. Methuen, London.

Cox, D.R. and Oakes, D. (1984). *Analysis of Survival Data*. Chapman and Hall, London.

Cox, D.R. and Snell, E.J. (1968). A general definition of residuals (with Discussion). *J. R. Stat. Soc. B*, **30**, 248–275.

Cox, D.R. and Snell, E.J. (1971). *Applied Statistics*. Methuen, London.

Cozzolino, J.M. (1968). Probabilistic models of decreasing failure rate processes. *Nav. Res. Logist. Q.*, **15**, 361–374.

Crellin, G.L., Jacobs, F.M., Smith, A.M., and Worledge, D.M. (1988). Organising dependent event data — a classification and analysis of multiple component fault reports. *Reliab. Eng.*, **15**, 145–158.

Crow, L.H. (1974). Reliability analysis for complex repairable systems. In *Reliability and Biometry* (ed. F. Proschan and R.J. Serfling), 379–410. SIAM, Philadelphia, PA.

Crow, L.H. (1982). Confidence interval procedures for the Weibull process with application to reliability growth. *Technometrics*, **24**, 251–256.

Crowder, M.J. (1985). A distribution model for repeated failure time measurements. *J. R. Stat. Soc. B*, **47**, 447–452.

Crowder, M.J., Kimber, A.C., Smith, R.L., and Sweeting, T.J. (1991). *Statistical Analysis of Reliability Data*. Chapman and Hall, London.

D'Agostino, R.B. and Stephens, M.A. (ed.) (1986). *Goodness-of-fit Techniques*. Marcel Dekker, New York, NY.

Dale, C.J. (1983). Application of proportional hazards in the reliability field. In *Proc. 4th Natl. Reliab. Conf., Birmingham*, 5B/1/1–9. National Centre of Systems Reliability, UKAEA, Warrington.

Davis, T.P. (1991). Competing risk survival analysis — theory and industrial applications. *Ph.D. thesis*, School of Mathematics and Statistics, University of Birmingham.

Dhillon, B.S. (1977). Literature survey on three-state device reliability systems. *Microelectron. Reliab.*, **16**, 601–602.

Downton, F. (1970). Bivariate exponential distributions in reliability theory. *J. R. Stat. Soc. B*, **32**, 408–417.

Drenick, R.F. (1960). The failure law of complex equipment. *J. Soc. Ind. Appl. Math.*, **8**, 680–690.

Drury, M., Walker, E.V., Wightman, D.M., and Bendell, A. (1987). Proportional hazards modelling in the analysis of computer systems reliability. In *Proc. 6th Natl. Reliab. Conf., Birmingham*, 5B/1/1–12. National Centre of Systems Reliability, UKAEA, Warrington.

Duane, J.T. (1964). Learning curve approach to reliability monitoring. *IEEE Trans. Aerosp.*, **2**, 563–566.

Edwards, G.T. (1988). A dependent failure study. *SRS Q. Dig.*, 9–17.

Edwards, G.T. (1989). Dependent failure assessment principles, engineering structure and data requirements. *Report SRS/GR/83*, National Centre of Systems Reliability, UKAEA, Warrington.

Elsayed, E.A. and Chan, C.K. (1990). Estimation of thin-oxide reliability using proportional hazards models. *IEEE Trans. Reliab.*, **39**, 329–335.

Etezardi-Amoli, J. and Ciampi, A. (1987). Extended hazard regression for censored survival data with covariates: a spline approximation for the baseline hazard function. *Biometrics*, **43**, 181–192.

Farewell, V.T. and Prentice, R.L. (1977). A study of distributional shape in lifetesting. *Technometrics*, **19**, 69–76.

Feller, W. (1968). *An Introduction to Probability Theory and its Applications, Vol. I*. John Wiley, New York, NY.

Fiegl, P. and Zelen, M. (1965). Estimation of exponential survival probabilities with concomitant information. *Biometrics*, **21**, 826–838.

Fleming, K.N. (1975). A reliability model for common mode failures in redundant safety systems. General Atomic Report No. GA-A13284.

Fleming, K.N., Mosleh, A., and Kelley, A.P. Jr (1983). Analysis of dependent failures in risk assessment and reliability evaluation. *Nucl. Saf.*, **24**, 637–657.

Forman, E.H. and Singpurwalla, N.D. (1977). An empirical stopping rule for debugging and testing computer software. *J. Am. Stat. Assoc.*, **72**, 750–757.

Freund, J.E. (1961). A bivariate extension of the exponential distribution. *J. Am. Stat. Assoc.*, **56**, 971–977.

Fussel, J.B. (1976). Fault tree analysis: concepts and techniques. In *Generic Techniques in Systems Reliability*, 133–162. NATO Advanced Institute, Noordhoof-Leyden.

Gail, M.H. (1975). A review and critique of some models used in competing risk analysis. *Biometrics*, **31**, 209–222.

Games, A.M., Amendola, A., and Martin, P. (1985). Multiple related failure events — risk, design, maintenance and cost. In *Proc. 5th Natl. Reliab. Conf., Birmingham*, 5B/4/1–7. National Centre of Systems Reliability, UKAEA, Warrington.

Games, A.M. and Ballard, G. (1988). A trigger-coupling mechanism model for dependent failures. In *Proc. 10th Adv. Reliab. Tech. Symp., Bradford* (ed. G.P. Libberton), 303–315. Elsevier Applied Science, London.

Glasser, G.L. (1967). The age replacement problem. *Technometrics*, **9**, 83–91.

Glasser, M. (1967). Exponential survival with covariance. *J. Am. Stat. Assoc.*, **62**, 561–568.

Gore, S.M., Pocock, S.J., and Kerr, G.R. (1984). Regression models and non-proportional hazards in the analysis of breast cancer survival. *Appl. Stat.*, **33**, 176–195.

Grigelionis, B.I. (1964). Limit theorems for sums of renewal processes. *Cybernetics in the Service of Communism, vol. 2*. In *Reliability Theory and Queueing Theory* (ed. A.I. Berg, N.G. Bruevich, and B.V. Gnedenko), 246–266. Energy Publishing House, Moscow.

Grove, D.M. and Davis, T.P. (1992). *Engineering, Quality, and Experimental Design, including Aspects of Taguchi*. Longman, Harlow.

Gumbel, E.J. (1960). Bivariate exponential distribution. *J. Am. Stat. Assoc.*, **55**, 698–707.

Halliday, B.R. and Devereux, I.F. (1983). Managing reliability growth in practice. In *Proc. Natl. Reliab. Conf., Birmingham*, 4A/1/1–9. National Centre of Systems Reliability, UKAEA, Warrington.

Holford, T.R. (1976). Life tables with concomitant information. *Biometrics*, **32**, 587–598.

Hughes, R.P. (1988). A framework for dependent failure analysis. In *Proc. 10th Adv. Reliab. Tech. Symp., Bradford* (ed. G.P. Libberton), 252–262. Elsevier Applied Science, London.

Hughes, R.P. (1989). Distributed failure probability approach to dependent failure analysis and its applications. *EureData, Sienna* (ed. V. Colombari), 167–177. Springer-Verlag, Berlin.

Humphreys, P., Games, A.M., Smith, A.M., and Worledge, D.M. (1987). Progress towards a better understanding of dependent failures by data collection, classification and improved modelling techniques. In *Proc. 6th Natl. Reliab. Conf., Birmingham.*, 2C/4/1–14. National Centre of Systems Reliability, UKAEA, Warrington.

Jack, N. and Dagpunar, J.S. (1992). Costing minimal-repair replacement policies over finite time horizons. *IMA J. Math. Appl. Bus. Ind.*, **3**, 207–217.

Jardine, A.K.S. and Anderson, M. (1984). Use of concomitant variables for reliability estimation and setting component replacement polices. In *Proc. 8th Adv. Reliab. Tech. Symp., Bradford*, B3/2/1–6. National Centre of Systems Reliability, UKAEA, Warrington.

Jelinski, Z. and Moranda, P.B. (1972). Software reliability research. In *Statistical Computer Performance Evaluations* (ed. W. Freiberger), 465–484. Academic Press, New York, NY.

Jenney, B.W. and Sherwin, D.J. (1986). Open and short circuit reliability of systems of identical items. *IEEE Trans. Reliab.*, **35**, 532–538.

Jewell, W.S. (1978a). Reliability growth as an artifact of renewal testing. *Operations Research Center Report ORC 78-9*, University of California, Berkeley, San Francisco, CA.

Jewell, W.S. (1978b). A curious renewal process average. *Operations Research Center Report ORC 78-12*, University of California, Berkeley, San Francisco, CA.

Johnson, N.L. and Kotz, S. (1972). *Distributions in Statistics: Continuous Multivariate Distributions.* John Wiley, New York, NY.

Kalbfleisch, J.D. (1974). Some efficiency calculations for survival distributions. *Biometrika*, **61**, 31–38.

Kalbfleisch, J.D., Peterson, A.V., Flournoy, N., Farewell, V.T., and Breslow, N.E. (1978). The analysis of failure time in the presence of competing risks. *Biometrics*, **34**, 541–554.

Kaplan, E.L. and Meier, P. (1958). Nonparametric estimation from incomplete observations. *J. Am. Stat. Assoc.*, **53**, 457–481.

Kay, R. (1977). Proportional hazards regression models and analysis of censored survival data. *Appl. Stat.*, **26**, 227–237.

Keating, J.P., Glaser, R.E., and Ketchum, N.S. (1990). Testing hypotheses about the shape parameter of a gamma distribution. *Technometrics*, **32**, 67–82.

Khintchine, A.Y. (1960). *Mathematical Methods in the Theory of Queueing.* Charles Griffin, London.

Kim, J.S. and Proschan, F. (1991). Piecewise exponential estimator of the survivor function. *IEEE Trans. Reliab.*, **40**, 134–139.

Kimber, A.C. (1990). Exploratory data analysis for possibly censored data from skewed distributions. *Appl. Stat.*, **39**, 21–30.

Kopocinski, B. (1974). Some estimations of two-terminal series-parallel system reliability. *Zastow. Matem.*, **14**, 9–16.

Koziol, J.A. and Green, S.B. (1976). A Cramer-von Mises statistic for randomly censored data. *Biometrika*, **63**, 465–474.

Lawless, J.F. (1982). *Statistical Models and Methods for Lifetime Data.* John Wiley, New York, NY.

Lawless, J.F. (1989). Contribution to the Discussion of Ansell and Phillips (1989). *Appl. Stat.*, **38**, 236–237.

Leach, C. (1988). Comparative reliability of some common design and specification methods. In *10th Adv. in Reliab. Tech. Symp., Bradford* (ed. G.P. Libberton), 247–251. Elsevier Applied Science, London.

Lee, L. (1980). Comparing rates of several independent Weibull processes. *Technometrics*, **22**, 427–430.

Lesanovsky, A. (1993). Systems with two dual failure modes — a survey. *Microelectron. Reliab.*, **33**, 1597–1626.

Lewis, P.A.W. (1964). A branching Poisson process model for the analysis of computer failure patterns. *J. R. Stat. Soc. B*, **26**, 398–456.

Lewis, P.A.W. (1967). Nonhomogeneous branching Poisson processes. *J. R. Stat. Soc. B*, **29**, 343–354.

Lewis, P.A.W. (1972). Recent results in the statistical analysis of univariate point processes. In *Stochastic Point Processes* (ed. P.A.W. Lewis), 1–54. Wiley-Interscience, New York, NY.

Lewis, P.A.W. and Robinson, D.W. (1974). Testing for a monotone trend in modulated renewal process. In *Reliability and Biometry* (ed. F. Proschan and R.J. Serfling), 163–182. SIAM, Philadelphia, PA.

Lewis, P.A.W. and Shedler, G.S. (1976). Statistical analysis of nonstationary series of events in a data base system. *IBM J. Res. Devl.*, 465–482.

Littlewood, B. (1981). Stochastic reliability growth: a model for fault removal in computer-programs and hardware-design. *IEEE Trans. Reliab.*, **30**, 313–320.

Littlewood, B. and Verrall, J.L. (1973). A Bayesian reliability growth model for computer software. *Appl. Stat.*, **22**, 332–346.

Lloyd, D.K. and Lipow, M. (1962). *Reliability: Management Methods and Mathematics*. Prentice Hall, London.

Lomnicki, Z.A. (1972). Two-terminal series-parallel networks. *Adv. Appl. Probab.*, **4**, 109–150.

Lomnicki, Z.A. (1973). Some aspects of the statistical approach to reliability (with Discussion). *J. R. Stat. Soc. A*, **136**, 395–420.

Makis, V. and Jardine, A.K.S. (1992). Computation of optimal policies in replacement models. *IMA J. Math. Appl. Bus. Ind.*, **3**, 169–175.

Malon, D.M. (1989). On a common error in open and short circuit reliability computation. *IEEE Trans. Reliab.*, **38**, 275–276.

Mann, N. (1945). Nonparametric test against trend. *Econometrika*, **13**, 245–259.

Mantel, N. (1966). Evaluation of survival data and two new rank order statistics arising from its consideration. *Cancer Chemother. Rep.*, **50**, 163–170.

Marshall, A.W. and Olkin, I. (1967). A multivariate exponential distribution. *J. Am. Stat. Assoc.*, **62**, 30–44.

McCullagh, P. and Nelder, J.A. (1989). *Generalized Linear Models*, 2nd edition. Chapman and Hall, London.

MIL-HDBK-189. (1981). *Reliability Growth Management*. Headquarters, US Army Communications Research and Development Command, Fort Monmouth, NJ.

Moore, E.F. and Shannon, C.E (1956). Reliable circuits using less reliable relays. *J. Franklin Inst.*, **262**, 191–208, 281–297.

Moss, T.R. (1991). Uncertainties in reliability statistics. *Reliab. Eng.*, **34**, 79–90.

Musa, J.D. (1975). A theory of software reliability and its application. *IEEE Trans. Software Eng.*, **1**, 312–327.

Musa, J.D. (1980). The measurement and management of software reliability. *Proc. IEEE*, **68**, 1131–1143.

Musa, J.D., Iannino, A., and Okumoto, K. (1987). *Software Reliability: Measurement, Prediction, Application*. McGraw-Hill, New York, NY.

Nakagawa, Y. and Hattori, Y. (1980). Reliability of all possible series-parallel redundant structures of m i.i.d. units with two failure modes. *IEEE Trans. Reliab.*, **29**, 320–322.

Nelson, W. (1969). Hazard plotting for incomplete failure data. *J. Qual. Tech.*, **1**, 27–52.

Nelson, W. (1988). Graphical analysis of system repair data. *J. Qual. Tech.*, **20**, 24–35.

Nelson, W. (1990). Hazard plotting of left truncated life data. *J. Qual. Tech.*, **22**, 230–238.

Nelson, W. (1993). *Accelerated Life Testing*. John Wiley, New York, NY.

Nelson, W. and Hahn, G.J. (1972). Linear estimation of a regression relationship from censored data. Part 1 — simple methods and their applications. *Technometrics*, **14**, 247–276.

Newton, D.W. (1991). Some pitfalls in reliability data analysis. *Reliab. Eng.*, **34**, 7–21.

O'Connor, P.D.T. (1981). *Practical Reliability Engineering*. Heyden, London.

O'Connor, P.D.T. (1991). Statistics in quality and reliability. *Reliab. Eng.*, **34**, 23–33.

Oliver, R.M. and Smith, J.Q. (1990). Influence diagrams, belief nets and decision analysis. In *Proc. Influence Diagrams for Inference, Prediction and Decision Making, Berkeley, California, 1988*. John Wiley, New York, NY.

Osaki, S. (1985). *Stochastic Systems Reliability Modelling*. World Scientific Publishers, Singapore.

Page, L.B. and Perry, J.E. (1988). Optimal series-parallel networks of 3-state devices. *IEEE Trans. Reliab.*, **37**, 388–394.

Perera, U.D. (1990). Reliability growth models and applications to a processor. In *Proc. 11th Adv. Reliab. Tech. Symp., Liverpool* (ed. P. Comer), 86–102. Elsevier Applied Science, London.

Peto, R. (1972). Contribution to the Discussion of Cox (1972). *J. R. Stat. Soc. B.*, **34**, 205–207.

Phillips, M.J. (1976). The reliability of two-terminal parallel-series networks subject to two kinds of failure. *Microelectron. and Reliab.*, **15**, 535–549.

Phillips, M.J. (1980). k-out-of-n:G systems are preferable. *IEEE Trans. Reliab.*, **29**, 166–169.

Phillips, M.J. (1981). A preventative maintenance plan for a system subject to revealed and unrevealed faults. *Reliab. Eng.*, **2**, 221–231.

Price, H.W. (1960). Reliability of parallel electronic components. *IRE Trans. Reliab. Qual. Control*, **9**, 35–39.

Proschan, F. and Sullo, P. (1976). Estimating the parameters of a multivariate exponential distribution. *J. Am. Stat. Assoc.*, **71**, 465–472.

Provan, J.S. and Ball, M.O. (1984). Computing network reliability in time polynomial in the number of cuts. *Oper. Res.*, **32**, 516–526.

Ripley, B. (1977). *Stochastic Simulation*. John Wiley, New York, NY.

Roger, J.H. (1985). Using factors when fitting the scale parameter to Weibull and other survival regression models with censored data. *GLIM Newsletter*, **11**, 14–15.

Rosner, N. (1961). Systems analysis — non-linear estimation techniques. In *Proc. Natl. Symp. on Reliab. and Qual. Control, New York, NY*, 203–207. IEEE, New York, NY.

Ross, S.M. (1970). *Applied probability models with optimization applications.* Holden-Day, San Francisco, CA.

Satoh, N., Sasaki, M., Yuge, T., and Yanagi, S. (1993). Reliability of 3-state device systems with simultaneous failures. *IEEE Trans. Reliab.*, **35**, 532–538.

Schoenfeld, D. (1982). Partial residuals for the proportional hazards regression model. *Biometrika*, **69**, 239–241.

Shaked, M. (1982). A general theory of some positive dependence notions. *J. Mult. Anal.*, **12**, 199–218.

Sheppard, J.M. (1983). Are reliability growth models practical planning and monitoring tools? In *Proc. 4th Natl. Reliab. Conf., Birmingham*, 4A/2/1–7. National Centre of Systems Reliability, UKAEA, Warrington.

Sherif, T.S. and Smith, M.L. (1981). Optimal maintenance models for systems subject to failure — a review. *Nav. Res. Logist. Q.*, **28**, 47–74.

Shier, D.R. and Liu, N. (1992). Bounding the reliability of networks. *J. Oper. Res. Soc.*, **43**, 539–548.

Smith, D.J. (1985). *Reliability and maintainability in perspective: practice, contractual, commercial and software aspects.* Macmillan, Software Reliability Library, London.

Smith, R.L. (1991). Weibull regression models for reliability analysis. *Reliab. Eng.*, **34**, 55–77.

Solomon, P.J. (1984). Effect of misspecification of regression models in the analysis of survival data. *Biometrika*, **71**, 291–298.

Soyer, R. (1986). Application of time series models to software reliability analysis. In *State of the Art Report on Software Reliability* (ed. A. Bendell and P. Mellor), 197–208. Pergamon Infotech, Maidenhead.

Stephens, M.A. (1989). Contribution to the Discussion of Ansell and Phillips (1989). *Appl. Stat.*, **38**, 235–236.

Sweeting, T.J. (1989). Contribution to the Discussion of Ansell and Phillips (1989). *Appl. Stat.*, **38**, 234–235.

Sweeting, T.J. (1992). Parameter-based asymptotics. *Biometrika*, **79**, 219–230.

Therneau, T.M., Grambsch, P.M., and Fleming, T.R. (1990). Martingale-based residuals for survival models. *Biometrika*, **77**, 147–160.

Thomas, L.C. (1986). A survey of maintenance and replacement models for maintainability and reliability of multi-item systems. *Reliab. Eng.*, **16**, 297–309.

Triner, D.A. (1986). The assessment of fleet equipment reliability. *Reliab. Eng.*, **14**, 63–74.

Triner, D.A. and Phillips, M.J. (1986). The reliability of equipment fitted to a fleet of ships. In *9th Adv. Reliab. Tech. Symp., Bradford*, C2/1/1–9. National Centre of Systems Reliability, UKAEA, Warrington.

Tukey, J. (1977). *Exploratory Data Analysis.* Addison-Wesley, Reading, MA.

Vesely, W.E. (1977). Estimating common cause failure probabilities in reliability and risk analysis: Marshall-Olkin specializations. In *Nuclear Systems Reliability Engineering and Risk Assessment* (ed. J.B. Fussel and G.R. Burdick), 314–341. SIAM, Philadelphia, PA.

von Alven, W.H. (ed.) (1964). *Reliability engineering.* Prentice Hall, Englewood Cliffs, NJ.

von Neumann, J. (1956). Probabilistic logics and the synthesis of reliable organisms from unreliable components. In *Automata Studies (Annals of Mathematical Studies, No. 34)*, (ed. C.E. Shannon and J. McCarthy), 43–98. Princeton University Press, Princeton, NJ.

Walls, L.A. (1994). A graphical approach to identification of dependent failures. *Research Report*, University of Paisley.

Walls, L.A. and Bendell, A. (1986). Time series methods in reliability. In *9th Adv. Reliab. Tech. Symp., Bradford*, C2/3/1–8. National Centre of Systems Reliability, UKAEA, Warrington.

Walls, L.A. and Bendell, A. (1989). Exploring field reliability data for potential dependent failures. In *Proc. Reliab. '89, Brighton*, 4Ab/3/1–8. National Centre of Systems Reliability, UKAEA, Warrington.

Weibull, W. (1951). A statistical distribution function of wide applicability. *J. Appl. Mech.*, **18**, 293–297.

Whitehead, J. (1980). Fitting Cox's regression model to survival data using GLIM. *Appl. Stat.*, **29**, 268–275.

Winfield, D.J. (1988). Long term reliability analysis of standby diesel generators. *Reliab. Eng.*, **21**, 293–308.

Zelen, M. (1959). Factorial experiments in life testing. *Technometrics*, **1**, 269–288.

Author index

Index